VOLUME THREE HUNDRED AND SIXTY NINE

INTERNATIONAL REVIEW OF CELL AND MOLECULAR BIOLOGY

New Antibody Formats

INTERNATIONAL REVIEW OF CELL AND MOLECULAR BIOLOGY

VOLUME THREE HUNDRED AND SIXTY NINE

INTERNATIONAL REVIEW OF CELL AND MOLECULAR BIOLOGY

New Antibody Formats

Edited by

FERNANDO ARANDA
Cima Universidad de Navarra, Pamplona, Spain

PEDRO BERRAONDO
Cima Universidad de Navarra, Pamplona, Spain

LORENZO GALLUZZI
Weill Cornell Medical College,
New York, NY, United States

ELSEVIER

ACADEMIC PRESS
An imprint of Elsevier

Academic Press is an imprint of Elsevier
50 Hampshire Street, 5th Floor, Cambridge, MA 02139, United States
525 B Street, Suite 1650, San Diego, CA 92101, United States
The Boulevard, Langford Lane, Kidlington, Oxford OX5 1GB, United Kingdom
125 London Wall, London, EC2Y 5AS, United Kingdom

First edition 2022

ISBN: 978-0-323-99402-6
ISSN: 1937-6448

Publisher: Zoe Kruze
Acquisitions Editor: Leticia M. Lima
Developmental Editor: Jhon Michael Peñano
Production Project Manager: Vijayaraj Purushothaman
Cover Designer: Matthew Limbert

Typeset by STRAIVE, India

Contents

Contributors

Luis Álvarez-Vallina
Cancer Immunotherapy Unit (UNICA), Department of Immunology, Hospital Universitario12 de Octubre; Immuno-Oncology and Immunotherapy Group, Instituto de Investigación Sanitaria 12 de Octubre (imas12), Madrid, Spain

Fernando Aranda
Program of Immunology and Immunotherapy, Cima Universidad de Navarra; Navarra Institute for Health Research (IDISNA), Pamplona, Spain

Leire Arrizabalaga
Program of Immunology and Immunotherapy, Cima Universidad de Navarra; Navarra Institute for Health Research (IDISNA), Pamplona, Spain

Robin Maximilian Awad
Laboratory for Molecular and Cellular Therapy, Department of Biomedical Sciences, Vrije Universiteit Brussel, Brussels, Belgium

Angela Bella
Program of Immunology and Immunotherapy, Cima Universidad de Navarra; Navarra Institute for Health Research (IDISNA), Pamplona, Spain

Pedro Berraondo
Program of Immunology and Immunotherapy, Cima Universidad de Navarra; Navarra Institute for Health Research (IDISNA), Pamplona; Centro de Investigación Biomédica en Red de Cáncer (CIBERONC), Madrid, Spain

Karine Breckpot
Laboratory for Molecular and Cellular Therapy, Department of Biomedical Sciences, Vrije Universiteit Brussel, Brussels, Belgium

Hannelore Ceuppens
Laboratory for Molecular and Cellular Therapy, Department of Biomedical Sciences, Vrije Universiteit Brussel, Brussels, Belgium

Assunta Cirella
Program of Immunology and Immunotherapy, Cima Universidad de Navarra; Navarra Institute for Health Research (IDISNA), Pamplona, Spain

Marta Compte
Department of Antibody Engineering, Leadartis S.L., Madrid, Spain

Matthias D'Huyvetter
In Vivo Cellular and Molecular Imaging Laboratory, Department of Medical Imaging, Vrije Universiteit Brussel, Brussels, Belgium

Timo W.M. De Groof
In Vivo Cellular and Molecular Imaging Laboratory, Department of Medical Imaging, Vrije Universiteit Brussel, Brussels, Belgium

Kim De Veirman
Laboratory for Hematology and Immunology, Department of Biomedical Sciences, Vrije Universiteit Brussel, Brussels, Belgium

Yannick De Vlaeminck
Laboratory for Molecular and Cellular Therapy, Department of Biomedical Sciences, Vrije Universiteit Brussel, Brussels, Belgium

Nick Devoogdt
In Vivo Cellular and Molecular Imaging Laboratory, Department of Medical Imaging, Vrije Universiteit Brussel, Brussels, Belgium

Claudia Augusta Di Trani
Program of Immunology and Immunotherapy, Cima Universidad de Navarra; Navarra Institute for Health Research (IDISNA), Pamplona, Spain

Thomas Ertveldt
Laboratory for Molecular and Cellular Therapy, Department of Biomedical Sciences, Vrije Universiteit Brussel, Brussels, Belgium

Myriam Fernandez-Sendin
Program of Immunology and Immunotherapy, Cima Universidad de Navarra; Navarra Institute for Health Research (IDISNA), Pamplona, Spain

Cleo Goyvaerts
Laboratory for Molecular and Cellular Therapy, Department of Biomedical Sciences, Vrije Universiteit Brussel, Brussels, Belgium

Stefan Haak
Pieris Pharmaceuticals GmbH, Hallbergmoos, Germany

Heleen Hanssens
In Vivo Cellular and Molecular Imaging Laboratory, Department of Medical Imaging, Vrije Universiteit Brussel, Brussels, Belgium

Sophie Hernot
In Vivo Cellular and Molecular Imaging Laboratory, Department of Medical Imaging, Vrije Universiteit Brussel, Brussels, Belgium

Shuyu Huang
Institute of Immunology and Infection Research, School of Biological Sciences, University of Edinburgh, Edinburgh, United Kingdom; Faculty of Veterinary Medicine, Department of Infectious Diseases and Immunology, Utrecht University, Utrecht, The Netherlands

Ahmet Krasniqi
In Vivo Cellular and Molecular Imaging Laboratory, Department of Medical Imaging, Vrije Universiteit Brussel, Brussels, Belgium

Quentin Lecocq
Laboratory for Molecular and Cellular Therapy, Department of Biomedical Sciences, Vrije Universiteit Brussel, Brussels, Belgium

Lukasz Mateusiak
In Vivo Cellular and Molecular Imaging Laboratory, Department of Medical Imaging, Vrije Universiteit Brussel, Brussels, Belgium

Fien Meeus
Laboratory for Molecular and Cellular Therapy, Department of Biomedical Sciences, Vrije Universiteit Brussel, Brussels, Belgium

Ignacio Melero
Program of Immunology and Immunotherapy, Cima Universidad de Navarra; Navarra Institute for Health Research (IDISNA); Department of Oncology, Clínica Universidad de Navarra, Pamplona; Centro de Investigación Biomédica en Red de Cáncer (CIBERONC), Madrid, Spain

Aizea Morales-Kastresana
Pieris Pharmaceuticals GmbH, Hallbergmoos, Germany

Vanessa Neiens
Pieris Pharmaceuticals GmbH, Hallbergmoos, Germany

Janet Peper-Gabriel
Pieris Pharmaceuticals GmbH, Hallbergmoos, Germany

Christine Rothe
Pieris Pharmaceuticals GmbH, Hallbergmoos, Germany

Laura Sanz
Molecular Immunology Unit, Hospital Universitario Puerta de Hierro Majadahonda, Majadahonda, Madrid, Spain

Aina Segués
Institute of Immunology and Infection Research, School of Biological Sciences, University of Edinburgh, Edinburgh, United Kingdom; Faculty of Veterinary Medicine, Department of Infectious Diseases and Immunology, Utrecht University, Utrecht, The Netherlands

Martin Siegemund
Pieris Pharmaceuticals GmbH, Hallbergmoos, Germany

Alice Sijts
Faculty of Veterinary Medicine, Department of Infectious Diseases and Immunology, Utrecht University, Utrecht, The Netherlands

Hana Valenta
Lab for Nanobiology, Department of Chemistry, KU Leuven, Leuven, Belgium

José Yélamos
Cancer Research Program, Hospital del Mar Medical Research Institute (IMIM), Unidad Asociada IIBB-CSIC; Immunology Unit, Department of Pathology, Hospital del Mar, Barcelona, Spain

Dietmar M. Zaiss
Institute of Immunology and Infection Research, School of Biological Sciences, University of Edinburgh, Edinburgh, United Kingdom; Department of Immune Medicine; Institute of Clinical Chemistry and Laboratory Medicine, University Hospital Regensburg; Institute of Pathology, University Regensburg, Regensburg, Germany

Katty Zeven
In Vivo Cellular and Molecular Imaging Laboratory, Department of Medical Imaging, Vrije Universiteit Brussel, Brussels, Belgium

CHAPTER ONE

Current innovative engineered antibodies

José Yélamos*
Cancer Research Program, Hospital del Mar Medical Research Institute (IMIM), Unidad Asociada IIBB-CSIC, Barcelona, Spain
Immunology Unit, Department of Pathology, Hospital del Mar, Barcelona, Spain
*Corresponding author: e-mail address: jyelamos@imim.es

Contents

Abstract

Antibody engineering has developed very intensively since the invention of the hybridoma technology in 1975, and it now can generate therapeutic agents with high specificity and reduced adverse effects. Indeed, antibodies have become one of the most innovative therapeutic agents in recent years, with some landing in the top 10 bestselling pharmaceutical drugs. New antibodies are being approved every year, in different formats and for treating various illnesses, including cancer, autoimmune inflammatory diseases, metabolic diseases and infectious diseases. In this review, I summarize current progress in innovative engineered antibodies. Overall, this progress has led to the approval by regulatory authorities of more than 100 antibody-based molecules, with many others at various stages of clinical development, indicating the high growth potential of the field.

International Review of Cell and Molecular Biology, Volume 369
ISSN 1937-6448
https://doi.org/10.1016/bs.ircmb.2022.03.007

1. Introduction

Antibodies are part of the adaptive immune system that are secreted by B cells after an antigen encounter, and they are able to recognize and interact with any substance that is foreign to the organism, such as viruses and bacteria. Human antibodies are large Y-shaped proteins that comprise four chains: two identical heavy (H) and two identical light (L) of linking protein domains. The light chain, of κ or λ, consists of a variable domain (VL) and a constant domain (CL); in turn, the heavy chains consist of a variable domain (VH) and three constant domains (CH1, CH2, and CH3) in IgA, IgD, and IgG, and four constant domains in IgE and IgM, whereby the variable domains are the ones that bind antigens (Chiu et al., 2019).

A large degree of diversity in the primary antibody repertoire is achieved by the distinct organization patterns of the genes encoding the variable regions of the antibody molecules, as the gene segments have to recombine with each other to form a functional gene that encodes for a particular antibody. Further diversity is generated by the random combination of heavy and light chain genes in each B cell, as well as by editing at the segment junctions (Hozumi and Tonegawa, 1976; Winter, 2019). Finally, an additional mechanism of diversity that occurs during the immune response is somatic hypermutation, which introduces point mutations in the rearranged variable region; this is followed by selection for B cells capable of producing antibodies with the highest affinity for the antigen, a process known as affinity maturation (Neuberger et al., 1998; Yélamos et al., 1995).

Since the high specificity and affinity of antibodies became known, the search for strategies to develop them as new therapeutic tools (human-made antibodies) (Winter and Milstein, 1991) in different pathological situations began. In this review, I summarize the approaches that have made possible to generate engineered antibodies with the desired specificity that are suitable for clinical use. Indeed, the number of engineered antibodies that have received approval by a regulatory authority, such as the U.S. Food and Drug Administration (FDA) or the European Medicines Agency (EMA), has been rising yearly (Fig. 1A), with some making it into the top 10 bestselling pharmaceutical drugs (Mullard, 2021). Of note, antibody-based therapies are already currently being used for numerous pathologies (Fig. 1B and C), and the existence of numerous clinical trials further indicates their enormous potential.

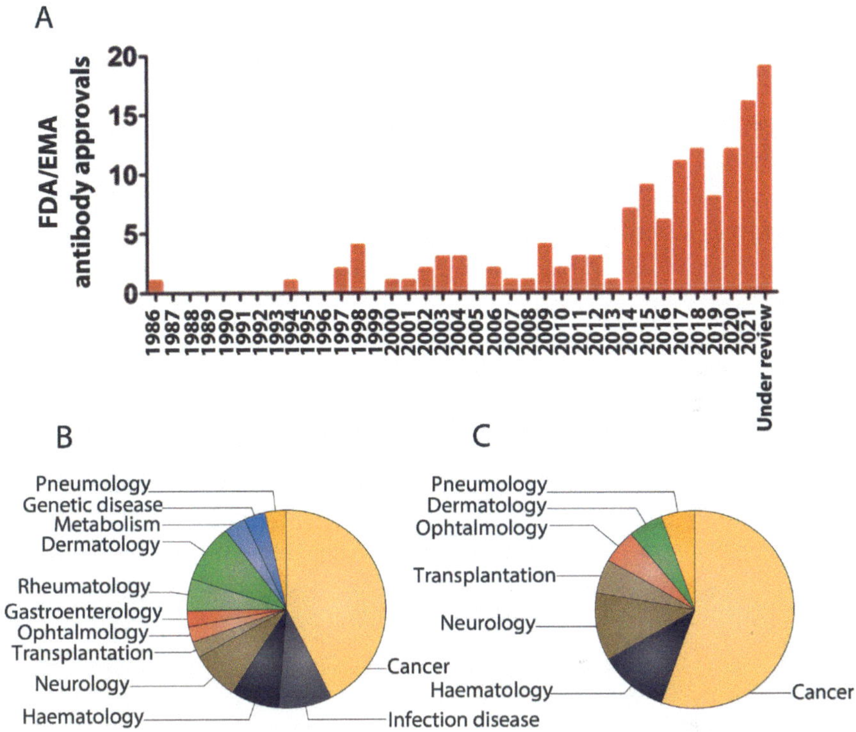

Fig. 1 Therapeutic antibodies approved or under review by the US Food and Drug Administration (FDA) and/or the European Medicines Agency (EMA). (A) Schematic representation of the evolution of the number of approved or under review antibodies. (B and C) Distribution of approved (B) or under review (C) antibodies by therapeutic areas. These areas are based on the indication of the first approval only. *Source: The Antibody Society (www.antibodysociety.org/antibody-therapeutics-product-data), accesed on December 2021.*

2. Historical overview of antibody engineering

Antibody engineering was developed starting with the invention of hybridoma technology by Georges Köhler and César Milstein in 1975 (Köhler and Milstein, 1975). In this original method, mice are immunized with an antigen, splenocytes are harvested, and the antigen-responding B-cells are immortalized by fusion with a myeloma cell line. Hybridomas are then screened to identify those that produce monoclonal antibodies for the antigen of interest. However, the animal origin of these antibodies limits their clinical application, as they themselves are immunogenic when administered to patients. Furthermore, these nonhuman antibodies interact poorly with the human immune effector system (Suzuki et al., 2015). Despite these limitations, some mouse monoclonal antibodies generated by this technology have been approved by the FDA and the EMA for certain pathological conditions (Table 1). Indeed, the first FDA-approved antibody for therapeutic use in 1986 was muromonab-CD3 (Orthoclone OKT3®), a murine anti-CD3 monoclonal antibody developed using the hybridoma technology to control acute rejection in organ transplantation (Emmons and Hunsicker, 1987; Smith, 1996). This antibody was finally withdrawn from the market in 2011.

Following the development of the hybridoma technology, the search began for strategies to remove or greatly reduce the immunogenicity of the antibody molecule to an acceptable level for therapeutic purposes. Eventually, protein engineering was used for this, which gave rise to the field of antibody engineering (Fig. 2). The first antibody engineering approach was to replace the constant region domains of the mouse antibody molecule with its human counterpart while maintaining the mouse antigen-binding domains, thereby generating chimeric antibodies. Such antibodies preserve the antigen specificity of the original mouse antibody with a human origin of 70% or more and can trigger human effector functions (Boulianne et al., 1984; Morrison et al., 1984) (Fig. 3), including antibody-dependent cell-mediated cytotoxicity (ADCC) and complement-dependent cytotoxicity (CDC) (Page et al., 2014). Abciximab, an anti-GPIIb/IIIa antigen-binding fragment (Fab), was the first chimeric antibody approved by the FDA in 1994, to be used in preventing blood clots in angioplasty (Simoons et al., 1994). Since then, 11 chimeric antibodies have been approved for clinical use (Table 2).

The next breakthrough in antibody engineering was the generation of humanized antibodies, in which only the complementary-determining regions (CDR) of the original murine antibody are transplanted into a human framework antibody molecule; this method, developed by Gregory Winter's laboratory, allows a human origin of up to 95% of the antibody molecule, while maintaining target specificity (Fig. 3) (Jones et al., 1986). Daclizumab, an anti-CD25 full-length antibody, was the first humanized antibody approved by FDA in 1997 to be used to prevent kidney transplant rejection (Vincenti et al., 1998). To date, a total of 62 humanized antibodies have been approved for clinical use in different pathologies (Table 3).

The great challenge of obtaining fully human therapeutic antibodies (Fig. 3) was achieved by the development of the antibody phage display technology by Winter's laboratory (Winter et al., 1994). In this approach, human VH and VL segments representing the whole immunoglobulin

Table 1 Mouse antibodies approved or under review by the US Food and Drug Administration (FDA) and/or the European Medicines Agency (EMA).

Name	Brand name	Target	Specificity	Format	Indication first approved or reviewed	EMA	FDA
Muromonab-CD3	Orthoclone Okt3	CD3	Monospecific	mIgG2a	Reversal of kidney transplant rejection	1986[a]	1986[b]
Tositumomab-I131	Bexxar	CD20	Monospecific	mIgG2a	Non-Hodgkin Lymphoma	NA	2003[b]
Edrecolomab	Panorex	EpCAM	Monospecific	mIgG2a	Colon cancer	1995[a,b]	NA
Catumaxomab	Removab	EPCAM, CD3	Bispecific	mIgG2a/k and rIgG2b/λ Hybrid	Malignant ascites	2009[b]	NA
Ibritumomab tiuxetan	Zevalin	CD20	Monospecific	mIgG1	Non-Hodgkin lymphoma	2004	2002
Blinatumomab	Blincyto	CD19, CD3	Bispecific	Tandem scFv	Acute lymphoblastic leukemia	2015	2014
Moxetumomab pasudotox (moxetumomab pasudotox-tdfk)	Lumoxiti	CD22	Monospecific	mIgG1 dsFv fused with PE38 exotoxin	Hairy cell leukemia	2021	2018
Inolimomb	(Pending)	CD25	Monospecific	mIgG1	Graft vs. host disease	NA	In review
Omburtamab	(Pending)	B7-H3 (CD276)	Monospecific	mIgG1	Central nervous system/ leptomeningeal metastasis from neuroblastoma	In review	NA

[a]Country-specific approval.
[b]Withdrawn or marketing discontinued.
NA, not approved or in review in the EU; not approved or information on review status not available in US.
Source: The Antibody Society (www.antibodysociety.org/antibody-therapeutics-product-data). Accessed December 2021.

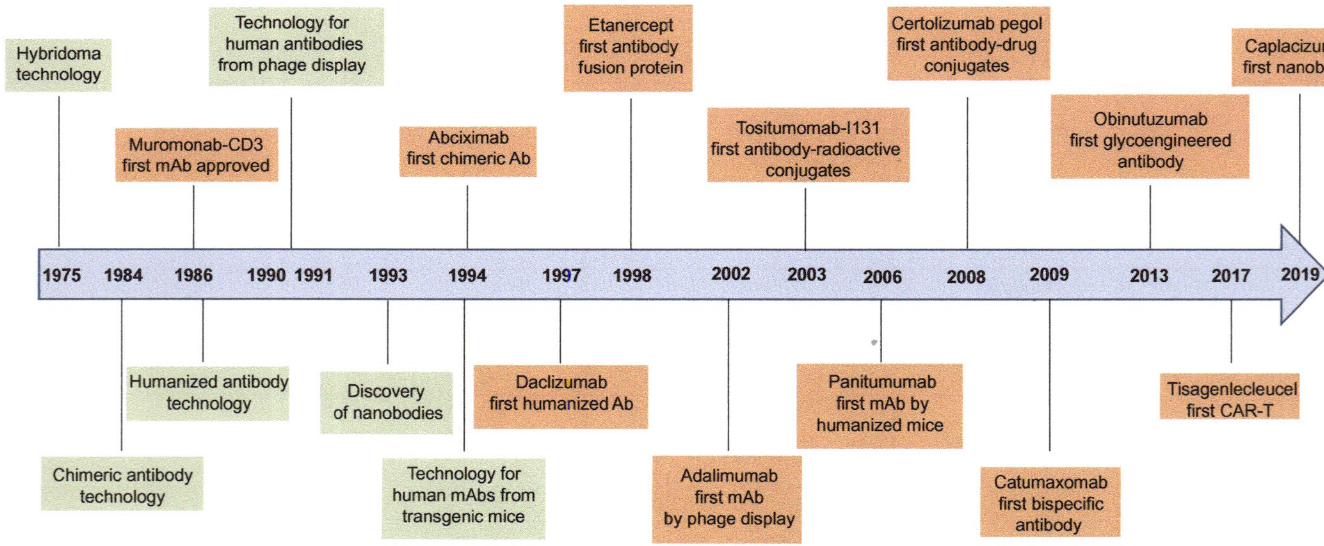

Fig. 2 Historical overview of antibody engineering.

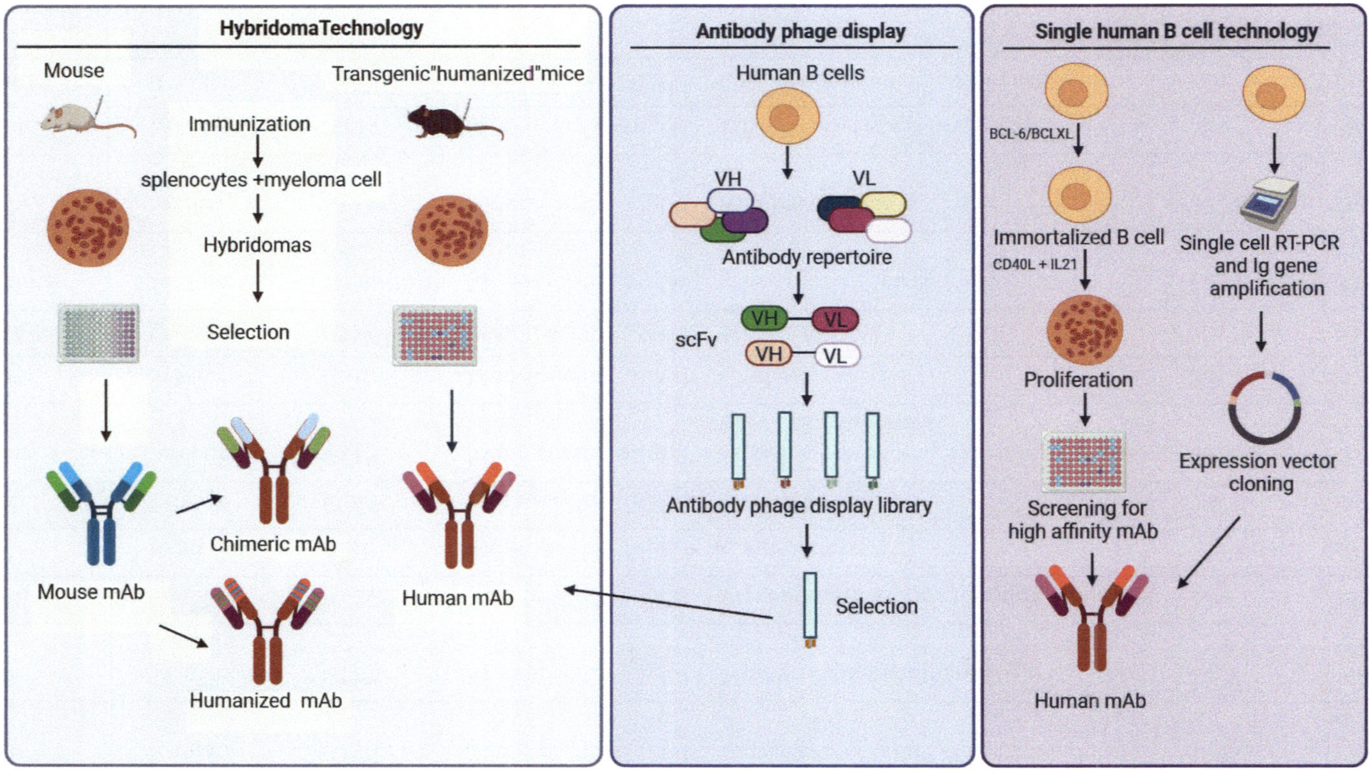

Fig. 3 Schematic overview of antibody engineering for the development of therapeutic antibodies. Image was created with BioRender.com.

Table 2 Chimeric antibodies approved or under review by the US Food and Drug Administration (FDA) and/or the European Medicines Agency (EMA).

Name	Brand name	Target	Specificity	Format	Indication first approved or reviewed	EMA	FDA
Abciximab	Reopro	GPIIb/IIIa	Monospecific	hIgG1 Fab	Prevention of blood clots in angioplasty	1995[a]	1994
Rituximab	MabThera, Rituxan	CD20	Monospecific	hIgG1	Non-Hodgkin lymphoma	1998	1997
Basiliximab	Simulect	IL-2R	Monospecific	hIgG1	Prevention of kidney transplant rejection	1998	1998
Infliximab	Remicade	TNF	Monospecific	hIgG1	Crohn disease	1999	1998
Cetuximab	Erbitux	EGFR	Monospecific	hIgG1	Colorectal cancer	2004	2004
Brentuximab vedotin	Adcetris	CD30	Monospecific	hIgG1	Hodgkin lymphoma, systemic anaplastic large cell lymphoma	2012	2011
Siltuximab	Sylvant	IL-6	Monospecific	hIgG1	Castleman disease	2014	2014
Dinutuximab	Unituxin	GD2	Monospecific	hIgG1	Neuroblastoma	2017; 2015[b]	2015
Obiltoxaximab	Anthim	B. anthrasis PA	Monospecific	hIgG1	Prevention of inhalational anthrax	2020	2016
Isatuximab (isatuximab-irfc)	Sarclisa	CD38	Monospecific	hIgG1	Multiple myeloma	2020	2020
Margetuximab-cmkb	MARGENZA	HER2	Monospecific	hIgG1	HER2+ metastatic breast cancer	NA	2020
Ublituximab	(Pending)	CD20	Monospecific	hIgG1	Chronic lymphocytic leukemia	NA	In review

[a]Country-specific approval.
[b]Withdrawn or marketing discontinued.
NA, not approved or in review in the EU; not approved or information on review status not available in US.
Source: The Antibody Society (www.antibodysociety.org/antibody-therapeutics-product-data). Accessed December 2021.

Table 3 Humanized antibodies approved or under review by the US Food and Drug Administration (FDA) and/or the European Medicines Agency (EMA).

Name	Brand name	Target	Specificity	Format	Indication first approved or reviewed	EMA	FDA
Efalizumab	Raptiva	CD11a	Monospecific	hIgG1	Psoriasis	2004[a]	2003[a]
Daclizumab	Zinbryta; Zenapax	CD25	Monospecific	hIgG1	Multiple sclerosis[a]; prevention of kidney transplant rejection[a]	2016[a]; 1999[a]	2016[a]; 1997[a]
Palivizumab	Synagis	RSV	Monospecific	hIgG1	Prevention of respiratory syncytial virus infection	1999	1998
Trastuzumab	Herceptin	HER2	Monospecific	hIgG1	Breast cancer	2000	1998
Omalizumab	Xolair	IgE	Monospecific	hIgG1	Asthma	2005	2003
Bevacizumab	Avastin	VEGF	Monospecific	hIgG1	Colorectal cancer	2005	2004
Natalizumab	Tysabri	a4 integrin	Monospecific	hIgG4	Multiple sclerosis	2006	2004
Ranibizumab	Lucentis	VEGF	Monospecific	hIgG1 Fab	Macular degeneration	2007	2006
Eculizumab	Soliris	C5	Monospecific	hIgG2(CH1-hinge)/ hIgG4(CH2-CH3)	Paroxysmal nocturnal hemoglobinuria	2007	2007
Certolizumab pegol	Cimzia	TNF	Monospecific	hFab	Crohn disease	2009	2008
Tocilizumab	RoActemra, Actemra	IL-6R	Monospecific	hIgG1	Rheumatoid arthritis	2009	2010
Pertuzumab	Perjeta	HER2	Monospecific	hIgG1	Breast Cancer	2013	2012

Continued

Table 3 Humanized antibodies approved or under review by the US Food and Drug Administration (FDA) and/or the European Medicines Agency (EMA).—cont'd

Name	Brand name	Target	Specificity	Format	Indication first approved or reviewed	EMA	FDA
Ado-trastuzumab emtansine	Kadcyla	HER2	Monospecific	hIgG1	Breast cancer	2013	2012
Obinutuzumab	Gazyva, Gazyvaro	CD20	Monospecific	hIgG1	Chronic lymphocytic leukemia	2014	2013
Vedolizumab	Entyvio	α4β7 integrin	Monospecific	hIgG1	Ulcerative colitis, Crohn disease	2014	2014
Pembrolizumab	Keytruda	PD1	Monospecific	hIgG4	Melanoma	2015	2014
Alemtuzumab	Lemtrada; MabCampath, Campath-1H	CD52	Monospecific	hIgG1	Multiple sclerosis; chronic myeloid leukemia[a]	2013; 2001[a]	2014; 2001[a]
Idarucizumab	Praxbind	Dabigatran	Monospecific	hIgG1 Fab	Reversal of dabigatran-induced anticoagulation	2015	2015
Mepolizumab	Nucala	IL-5	Monospecific	hIgG1	Severe eosinophilic asthma	2015	2015
Elotuzumab	Empliciti	SLAMF7	Monospecific	hIgG1	Multiple myeloma	2016	2015
Ixekizumab	Taltz	IL-17a	Monospecific	hIgG4	Psoriasis	2016	2016
Reslizumab	Cinqaero, Cinqair	IL-5	Monospecific	hIgG4	Asthma	2016	2016

Atezolizumab	Tecentriq	PD-L1	Monospecific	hIgG1	Bladder cancer	2017	2016
Inotuzumab ozogamicin	BESPONSA	CD22	Monospecific	hIgG4	Acute lymphoblastic leukemia	2017	2017
Emicizumab	Hemlibra	Factor Ixa, X	Bispecific	hIgG4	Hemophilia A	2018	2017
Ocrelizumab	OCREVUS	CD20	Monospecific	hIgG1	Multiple sclerosis	2018	2017
Benralizumab	Fasenra	IL-5Rα	Monospecific	hIgG1	Asthma	2018	2017
Gemtuzumab ozogamicin	Mylotarg	CD33	Monospecific	hIgG4	Acute myeloid leukemia	2018	2017; 2000[a]
Galcanezumab (galcanezumab-gnlm)	Emgality	CGRP	Monospecific	hIgG4	Migraine prevention	2018	2018
Mogamulizumab (mogamulizumab-kpkc)	Poteligeo	CCR4	Monospecific	hIgG1	Mycosis fungoides or Sézary syndrome	2018	2018
Tildrakizumab (tildrakizumab-asmn)	Ilumya	IL-23 p19	Monospecific	hIgG1	Plaque psoriasis	2018	2018
Fremanezumab (fremanezumab-vfrm)	Ajovy	CGRP	Monospecific	hIgG2Δa	Migraine prevention	2019	2018
Ravulizumab (ravulizumab-cwvz)	Ultomiris	C5	Monospecific	hIgG2(CH1-hinge)/ hIgG4(CH2-CH3)	Paroxysmal nocturnal hemoglobinuria	2019	2018
Ibalizumab (ibalizumab-uiyk)	Trogarzo	CD4	Monospecific	hIgG4	HIV infection	2019	2018

Continued

Table 3 Humanized antibodies approved or under review by the US Food and Drug Administration (FDA) and/or the European Medicines Agency (EMA).—cont'd

Name	Brand name	Target	Specificity	Format	Indication first approved or reviewed	EMA	FDA
Caplacizumab (caplacizumab-yhdp)	Cablivi	von Willebrand factor	Monospecific	Bivalent nanobody	Acquired thrombotic thrombocytopenic purpura	2018	2019
Risankizumab (risankizumab-rzaa)	Skyrizi	IL-23 p19	Monospecific	hIgG1	Plaque psoriasis	2019	2019
Polatuzumab vedotin (polatuzumab vedotin-piiq)	Polivy	CD79b	Monospecific	hIgG1	Diffuse large B-cell lymphoma	2020	2019
Romosozumab (romosozumab-aqqg)	Evenity	Sclerostin	Monospecific	hIgG2	Osteoporosis in postmenopausal women at increased risk of fracture	2019	2019
Brolucizumab (brolucizumab-dbll)	Beovu	VEGF-A	Monospecific	hscFv	Neovascular age-related macular degeneration	2020	2019
Crizanlizumab (crizanlizumab-tmca)	Adakveo	CD62 (aka P-selectin)	Monospecific	hIgG2	Sickle cell disease	2020	2019
[fam-]trastuzumab deruxtecan, (fam-trastuzumab deruxtecan-nxki)	Enhertu	HER2	Monospecific	hIgG1	HER2+ metastatic breast cancer	2021	2019

Eptinezumab (eptinezumab-jjmr)	VYEPTI	CGRP	Monospecific	hIgG1	Migraine prevention	EU review	2020
Sacituzumab govitecan (sacituzumab govitecan-hziy)	TRODELVY	TROP-2	Monospecific	hIgG1	Triple-neg. breast cancer	2021	2020
Inebilizumab (inebilizumab-cdon)	Uplizna	CD19	Monospecific	hIgG1	Neuromyelitis optica spectrum disorders	EU review	2020
Tafasitamab (tafasitamab-cxix)	Monjuvi, Minjuvi	CD19	Monospecific	hIgG1/2 hybrid	Diffuse large B-cell lymphoma	2021	2020
Belantamab mafodotin (belantamab mafodotin-blmf)	BLENREP	B-cell maturation antigen	Monospecific	hIgG1	Multiple myeloma	2020	2020
Satralizumab (satralizumab-mwge)	Enspryng	IL-6R	Monospecific	hIgG2	Neuromyelitis optica spectrum disorder	2021	2020
Naxitamab-gqgk	DANYELZA	GD2	Monospecific	hIgG1	High-risk neuroblastoma and refractory osteomedullary disease	NA	2020
Loncastuximab tesirine	Zynlonta	CD19	Monospecific	hIgG1/k	Diffuse large B-cell lymphoma	NA	2021
Dostarlimab	Jemerli	PD-1	Monospecific	hIgG4	Endometrial cancer	2021	2021

Continued

Table 3 Humanized antibodies approved or under review by the US Food and Drug Administration (FDA) and/or the European Medicines Agency (EMA).—cont'd

Name	Brand name	Target	Specificity	Format	Indication first approved or reviewed	EMA	FDA
Teplizumab	(Pending)	CD3	Monospecific	hIgG1	Type 1 diabetes	NA	In review (2nd cycle)
Retifanlimab	(Pending)	PD-1	Monospecific	hIgG4	Carcinoma of the anal canal	EU review	In review (2nd cycle)
Oportuzumab monatox	(Pending)	EpCAM	Monospecific	hscFv fused with PE38 exotoxin	Bladder cancer	NA	In review (2nd cycle)
Bimekizumab	Bimzelx	IL-17A and IL-17F (overlapping binding site)	Monospecific	hIgG1	Psoriasis	2021	In review
Sutimlimab	(Pending)	C1s	Monospecific	hIgG4	Cold agglutinin disease	NA	In review (2nd cycle)
Toripalimab	Tuoyi	PD-1	Monospecific	hIgG4	Nasopharyngeal carcinoma	NA	In review
Penpulimab	(Pending)	PD-1	Monospecific	hIgG1	Metastatic nasopharyngeal carcinoma	NA	In review
Faricimab	(Pending)	VEGF-A, Ang-2	Bispecific	hIgG1	Neovascular age-related macular degeneration, diabetic macular edema	EU review	In review

Tebentafusp	(Pending)	gp100, CD3	Bispecific	hscFV fused with a TCR	Metastatic uveal melanoma	EU review	In review
Tislelizumab	(Pending)	PD-1	Monospecific	hIgG4	Esophageal squamous cell carcinoma	NA	In review
Lecanemab	(Pending)	Amyloid beta protofibrils	Monospecific	hIgG1	Alzheimer's disease	NA	In review
Spesolimab	(Pending)	IL-36 receptor	Monospecific	hIgG1	Generalized pustular psoriasis	EU review	NA
Donanemab	(Pending)	Amyloid beta, N3pG (N-terminal truncated)	Monospecific	hIgG1	Alzheimer's disease	NA	In review

[a]Withdrawn or marketing discontinued.
NA, not approved or in review in the EU; not approved or information on review status not available in US.
Source: The Antibody Society (www.antibodysociety.org/antibody-therapeutics-product-data). Accessed December 2021.

gene-encoding repertoire are harvested from human B lymphocytes or assembled in vitro and then ligated into a phage display vector, thereby generating a single-chain fragment variable (Fv) or antigen-binding fragments (Fab) phage-display library. Phages are then selected from the repertoire by their binding to specific antigens, and the selected antibody fragments are secreted from infected bacteria. In addition, the affinity of the selected antibody can be improved by site-directed mutagenesis of CDR and affinity selection (Winter et al., 1994). The success of this technology was demonstrated with the development of adalimumab (Humira), the first fully human antibody approved by FDA in 2002, which is directed against the tumor necrosis factor α (TNFα) for treating rheumatoid arthritis (Den Broeder et al., 2002; Kempeni, 1999).

A second successful technical approach to obtain fully human therapeutic antibodies is to use the hybridoma technology in transgenic "humanized" mice models, in which the mouse immunoglobulin gene loci have been replaced with the human immunoglobulin gene loci (Lonberg, 2005; Mendez et al., 1997). Currently, there are seven major platforms for the generation of fully human antibodies in transgenic animals (Lu et al., 2020). The first human therapeutic antibody developed by one of these platforms was panitumumab (Vectibix), an anti-epidermal growth factor receptor (EGFR) antibody, approved by the FDA in 2006 for the treatment of colorectal cancer patients (Gibson et al., 2006; Moroni et al., 2005). To date, 47 fully human antibodies have been approved for clinical use by the FDA using either of the two approaches—antibody phage display or transgenic humanized mice models (Table 4).

Other methods to generate fully human antibodies are based on single, human B cell isolation (Fig. 3). In one of these approaches, the immunoglobulin heavy and light chain gene transcripts from each cell are amplified by nested RT-PCR and cloned into eukaryotic expression vectors to produce monoclonal human antibodies (Tiller et al., 2008). The second approach is to immortalize single human B cells by Epstein-Barr virus transformation (Corti and Lanzavecchia, 2013) or by expression of BCL-6 and expansion with CD40L and cytokines (Kwakkenbos et al., 2016).

After the successful completion of the first antibody engineering phase, which focused on generating human antibodies that can be administered for therapeutic use, antibody engineering has entered a very active phase of format modification (Fig. 4) that aims to increase the therapeutic effects of these antibodies (Fig. 5). In this way, antibody engineering is contributing to the development of numerous antibody-based products in the pharmaceutical market, with a high therapeutic interest.

Table 4 Fully human antibodies approved or under review by the US Food and Drug Administration (FDA) and/or the European Medicines Agency (EMA).

Name	Brand name	Target	Specificity	Backbone	Indication first approved or reviewed	EMA	FDA
Nebacumab	Centoxin	Endotoxin	Monospecific	hIgM	Gram-negative sepsis	1991[a,b]	NA
Adalimumab	Humira	TNF	Monospecific	hIgG1	Rheumatoid arthritis	2003	2002
Panitumumab	Vectibix	EGFR	Monospecific	hIgG2	Colorectal cancer	2007	2006
Ustekinumab	Stelara	IL-12/23	Monospecific	hIgG1	Psoriasis	2009	2009
Canakinumab	Ilaris	IL-1β	Monospecific	hIgG1	Muckle-Wells syndrome	2009	2009
Golimumab	Simponi	TNF	Monospecific	hIgG1	Rheumatoid and psoriatic arthritis, ankylosing spondylitis	2009	2009
Ofatumumab	Arzerra	CD20	Monospecific	hIgG1	Chronic lymphocytic leukemia	2010[a]	2009
Denosumab	Prolia	RANK-L	Monospecific	hIgG2	Bone loss	2010	2010
Belimumab	Benlysta	BLyS	Monospecific	hIgG1	Systemic lupus erythematosus	2011	2011
Ipilimumab	Yervoy	CTLA-4	Monospecific	hIgG1	Metastatic melanoma	2011	2011
Raxibacumab	(Pending)	B. anthrasis PA	Monospecific	hIgG1	Anthrax infection	NA	2012
Ramucirumab	Cyramza	VEGFR2	Monospecific	hIgG1	Gastric cancer	2014	2014
Nivolumab	Opdivo	PD1	Monospecific	hIgG4	Melanoma, non-small cell lung cancer	2015	2014
Evolocumab	Repatha	PCSK9	Monospecific	hIgG2	High cholesterol	2015	2015

Continued

Table 4 Fully human antibodies approved or under review by the US Food and Drug Administration (FDA) and/or the European Medicines Agency (EMA).—cont'd

Name	Brand name	Target	Specificity	Backbone	Indication first approved or reviewed	EMA	FDA
Necitumumab	Portrazza	EGFR	Monospecific	hIgG1	Non-small cell lung cancer	2015	2015
Secukinumab	Cosentyx	IL-17a	Monospecific	hIgG1	Psoriasis	2015	2015
Alirocumab	Praluent	PCSK9	Monospecific	hIgG1	High cholesterol	2015	2015
Daratumumab	Darzalex	CD38	Monospecific	hIgG1	Multiple myeloma	2016	2015
Olaratumab	Lartruvo	PDGFRα	Monospecific	hIgG1	Soft tissue sarcoma	2016[b]	2016
Bezlotoxumab	Zinplava	Clostridium difficile enterotoxin B	Monospecific	hIgG1	Prevention of Clostridium difficile infection recurrence	2017	2016
Brodalumab	Siliq, LUMICEF, Kyntheum	IL-17R	Monospecific	hIgG2	Plaque psoriasis	2017	2017
Dupilumab	Dupixent	IL-4Rα	Monospecific	hIgG4	Atopic dermatitis	2017	2017
Guselkumab	TREMFYA	IL-23 p19	Monospecific	hIgG1	Plaque psoriasis	2017	2017
Sarilumab	Kevzara	IL-6R	Monospecific	hIgG1	Rheumatoid arthritis	2017	2017
Avelumab	Bavencio	PD-L1	Monospecific	hIgG1	Merkel cell carcinoma	2017	2017
Durvalumab	IMFINZI	PD-L1	Monospecific	hIgG1	Bladder cancer	2018	2017

Erenumab (erenumab-aooe)	Aimovig	CGRP receptor	Monospecific	hIgG2	Migraine prevention	2018	2018
Burosumab (burosumab-twza)	Crysvita	FGF23	Monospecific	hIgG1	X-linked hypophosphatemia	2018	2018
Lanadelumab (lanadelumab-flyo)	Takhzyro	Plasma kallikrelin	Monospecific	hIgG1	Hereditary angioedema attacks	2018	2018
Cemiplimab (cemiplimab-rwlc)	Libtayo	PD-1	Monospecific	hIgG4	Cutaneous squamous cell carcinoma	2019	2018
Emapalumab (emapalumab-lzsg)	Gamifant	IFNg	Monospecific	hIgG1	Primary hemophagocytic lymphohistiocytosis	NA	2018
Enfortumab vedotin (enfortumab vedotin-ejfv)	Padcev	Nectin-4	Monospecific	hIgG1	Urothelial cancer	EU review	2019
Teprotumumab (teprotumumab-trbw)	Tepezza	IGF-1R	Monospecific	hIgG1	Thyroid eye disease	NA	2020
Atoltivimab, maftivimab, and odesivimab-ebgn	Inmazeb	Ebola virus	Monospecific	hIgG1	Ebola virus infection	NA	2020
Ansuvimab-zykl	Ebanga	Ebola virus glycoprotein	Monospecific	hIgG1	Ebola virus infection	NA	2020

Continued

Table 4 Fully human antibodies approved or under review by the US Food and Drug Administration (FDA) and/or the European Medicines Agency (EMA).—cont'd

Name	Brand name	Target	Specificity	Backbone	Indication first approved or reviewed	EMA	FDA
Evinacumab	Evkeeza	Angiopoietin-like 3	Monospecific	hIgG4	Homozygous familial hypercholesterolemia	2021	2021
Amivantamab	RYBREVANT	EGFR, cMET	Bispecific	hIgG1	NSCLC w/EGFR exon 20 insertion mutations	EU review	2021
Aducanumab	ADUHELM	Amyloid beta	Monospecific	hIgG1	Alzheimer's disease	EU review	2021
Tralokinumab	(Pending)	IL-13	Monospecific	hIgG4	Atopic dermatitis	2021	In review (2nd cycle)
Narsoplimab	(Pending)	MASP-2	Monospecific	hIgG4	Hematopoietic stem cell transplant-associated thrombotic microangiopathies	NA	In review
Anifrolumab, anifrolumab-fnia	Saphnelo	IFNAR1	Monospecific	hIgG1	Systemic lupus erythematosus	EU review	2021
Tisotumab vedotin, tisotumab vedotin-tftv	TIVDAK	Tissue factor	Monospecific	hIgG1	Cervical cancer	NA	2021

Sintilimab	Tyvyt	PD-1	Monospecific	hIgG4	Non-small cell lung cancer	NA	In review
Tezepelumab	(Pending)	Thymic stromal lymphopoietin	Monospecific	hIgG2	Severe asthma	EU review	In review
Relatlimab	(Pending)	LAG-3	Monospecific	hIgG4	Melanoma	NA	In review
Casirivimab + imdevimab	REGEN-COV, Ronapreve	SARS-CoV-2	Monospecific	hIgG1	COVID-19	2021	In review
Regdanvimab	(Pending)	SARS-CoV-2	Monospecific	hIgG1	COVID-19	2021	NA

[a]Country-specific approval.
[b]Withdrawn or marketing discontinued.
NA, not approved or in review in the EU; not approved or information on review status not available in US.
Source: The Antibody Society (www.antibodysociety.org/antibody-therapeutics-product-data). Accessed December 2021.

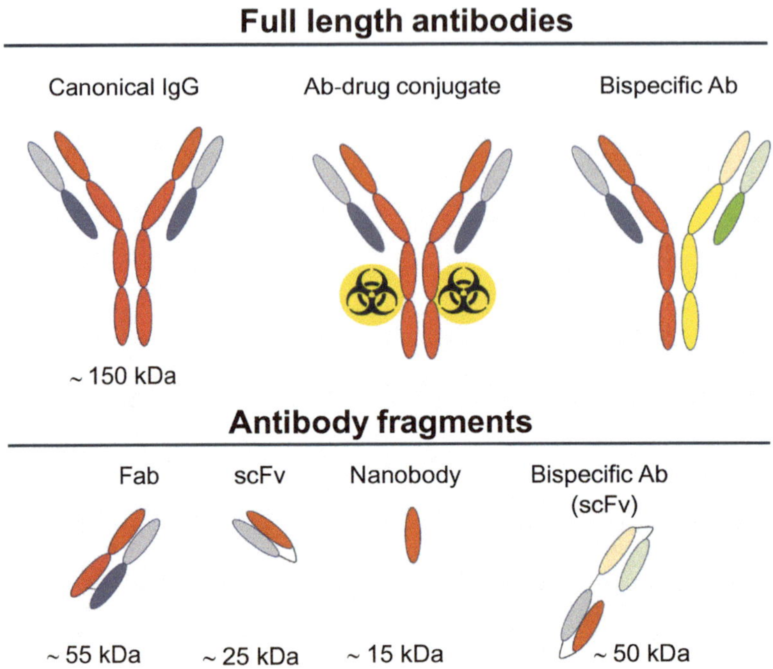

Fig. 4 Schematic representation of antibody formats. These formats can be further subcategorized. For example, there are around 100 different bispecific antibody formats.

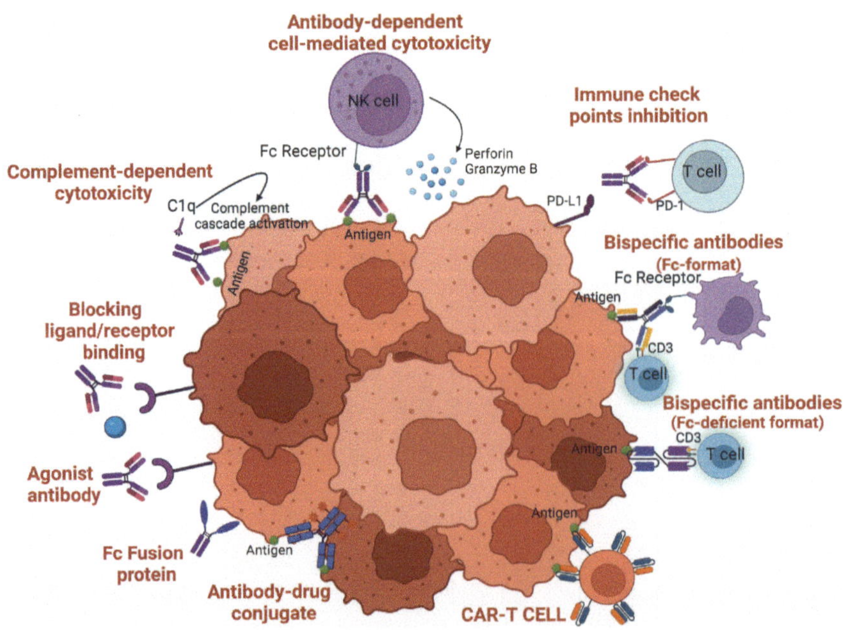

Fig. 5 Schematic representation of the mechanisms of action of antibody-based therapeutic in cancer treatment. Image was created with BioRender.com.

3. Bispecific antibodies

With the exception of IgG4, which can form hybrid molecules with specificity for two different antigens (Schuurman et al., 1999), all IgGs are monospecific and recognize a single antigen (Chiu et al., 2019). The idea of generating antibodies with two antigenic specificities is ancient (Nisonoff and Rivers, 1961) but could not be developed until the late 1990s, which was thanks to advances in antibody engineering (Merchant et al., 1998). Bispecific antibodies are designed to recognize two different antigens or two different epitopes in the same antigen simultaneously (Labrijn et al., 2019; Ma et al., 2021).

The great expansion of this area of antibody engineering is evidenced by the description of more than 100 different bispecific antibody formats to date (Brinkmann and Kontermann, 2017). The antigen-binding domains of bispecific antibodies can be arranged in different formats, including scFv, Fab, single domain antibody (sdAbs), and large IgG-like molecules (Blanco et al., 2021). Symmetric IgG-based bispecific antibodies comprise unmodified heavy chain constant regions, whereas asymmetric IgG-based bispecific antibodies comprise modified heavy chains to force hetero-dimerization using the knobs-into-holes strategy (Ridgway et al., 1996). Fc-less bispecific antibodies can be constructed by connecting two scFv through a linker (Brinkmann and Kontermann, 2017).

The mechanisms of action of bispecific antibodies include recruitment and activation of immune effector cells, blocking of immune check points, blocking of inflammatory factors, and blocking of dual signaling pathways (Blanco et al., 2021; Labrijn et al., 2019). Although the development of bispecific antibodies has been very much focused on cancer immunotherapy, there are other pathologies for which they may be useful, such as hemophilia A, neovascular age-related macular degeneration, diabetes, bacterial pneumonia, and Alzheimer's disease (Blanco et al., 2021; Ma et al., 2021). Catumaxomab was the first approved bispecific antibody in 2009, and it is directed against CD3 and EpCAM for treating solid tumors in patients with malignant ascites (Heiss et al., 2010). This antibody was withdrawn from the market in 2017 for commercial reasons. To date, three additional bispecific antibodies (blinatumomab, emicizumab and amivantamab) have been approved for clinical uses, one (faricimab) is currently under review (Tables 1–4), and more than 200 are at different stages of clinical development.

4. Nanobodies

Nanobodies, also termed sdAbs, are single-domain heavy chain-only antibodies derived from camelids (Muyldermans, 2021). Nanobodies have numerous beneficial properties, such as their small size that allow them easier access to tissues, high stability, and a strong affinity to their cognate antigen, all of which make them very good candidates for development into next-generation biopharmaceutical agents (Steeland et al., 2016). Several types of libraries are currently available to retrieve nanobodies with antigenic specificities of interest, for instance using phage display technology (Arbabi Ghahroudi et al., 1997). The selected nanobodies are then humanized (Vincke et al., 2009) and engineered into different formats, such as bispecific constructs (Chanier and Chames, 2019), radionucleotide-conjugated antibodies (Balhuizen et al., 2017), or as the antigenic recognition component of chimeric antigen receptor (CAR)-T cells (Xie et al., 2019, 2020).

Caplacizumab, a bivalent nanobody against the von Willebrand factor, was the first nanobody approved in 2019 for treating acquired thrombotic thrombocytopenic purpura, a rare disease characterized by excessive blood clotting in small blood vessels (Scully et al., 2019). Approval of caplacizumab has been an endorsement of the development of nanobodies for therapy, and there are currently 14 clinical trials with nanobody-based products.

5. Intrabodies

Intrabodies are directed toward intracellular targets in order to regulate allosteric proteins, manipulate protein functionality, and target degradation (Redchuk et al., 2020; Zhang et al., 2020). When provided with the corresponding protein trafficking signals, intrabodies can be directed to enter the endoplasmic reticulum via addition of a SEKDEL retention signal (Munro and Pelham, 1987), to enter the nucleus via the SV40-derived nuclear localization signal (Wheeler et al., 2003), or to remain in the cytoplasm simply by deletion of the leader peptide (Marschall et al., 2011).

Cytosolic intrabodies have been especially studied for targeting Ras (Quevedo et al., 2018; Tanaka et al., 2007). Intracellular expression of a neutralizing antibody specifically promotes apoptosis in vitro in human cancer cells, but not in untransformed cells. Moreover, intratumor transduction of colon carcinoma cells with the anti-Ras antibody, using an adenoviral vector, elicits sustained tumor regression in mice (Tanaka et al., 2007). The preferred targets for nuclear antibody are cell cycle proteins that are dysregulated in tumor cells (Chen et al., 2019; Wu et al., 2020). A nuclear-localizing form of an anti-c-Myb antibody blocks its transactivating activity in a dose-dependent manner, and was shown to inhibit proliferation of human leukemia cells (Kasono et al., 1998). Similarly, stable expression of a nuclear-targeted anti-cyclin E intrabody inhibits the growth of a breast cancer cell line (Strube and Chen, 2002). In addition, pre-clinical studies have shown that intrabodies inhibit β-catenin-mediated Wnt signaling (Newnham et al., 2015).

6. Fine-tuning Fc-dependent effector functions

Fc-dependent effector functions of antibodies (ADCC and CDC) may significantly affect their therapeutic effectiveness. ADCC occurs when an antibody binds through its variable region to a cognate antigen, and through its Fc domain to Fcγ receptors (FcγRs), and mainly FcγRIIIa, on the surface of an effector cell. CDC results from the C1q protein binding to the Fc domain of antibodies that are in turn bound to a cell surface antigen (Van Erp et al., 2019). In some situations, the aim of Fc modification is to increase binding to FcγRIIIa, and therefore to enhance the ADCC activity of the antibody (Barnhart and Quigley, 2017; Brezski and Georgiou, 2016; Sondermann and Szymkowski, 2016). However, in other situations, such as in the case of checkpoint inhibitors (e.g., PD1-1/PD-L1), ADCC induction can be detrimental, leading to destruction of the effector immune cell; in these cases, it is necessary to introduce changes in the Fc region that reduce or eliminate Fc effector functions (Borrok et al., 2017; Labrijn et al., 2008; Lo et al., 2017).

Tuning Fc-dependent effector functions has been achieved by amino acid changes in the Fc region, which can produce distinct effects; for instance, they can: (i) enhance ADCC by improving FcγRIIIa binding; (ii) remove Fc effector functions; (iii) increase the antibody half-life; and/or (iv) lead to hinge stabilization (Lazar et al., 2006; Shields et al., 2001; Vafa et al., 2014). A second approach to modulating Fc-mediated effector functions is to alter glycosylation patterns (glycoengineering) (Van Landuyt et al., 2019), such as by introducing low or no fucosylation, to improve FcγRIIIa binding (Golay et al., 2013; Shields et al., 2002). In addition, producing aglycosylated antibodies is also a strategy to remove Fc effector functions (Walker et al., 1989). Obinutuzumab (anti-CD20), the first glycoengineered antibody approved by FDA that is in a low fucose form, can activate neutrophils and mediate phagocytosis through CD16 more efficiently than rituximab (anti-CD20) (Golay et al., 2013). To date, 38 antibodies with modified Fc functionality have been approved; of these, nine were created by glycoengineering, and the remaining 29, by amino acid changes, highlighting the importance of these modifications in antibody engineering (Table 5).

Table 5 Therapeutic antibodies with modified Fc functionality approved by FDA and EMA.

Name	Brand name	Target	Fc modifications	Reason for Fc modifications	Indication first approved or reviewed	EMA	FDA
Obinutuzumab	Gazyva, Gazyvaro	CD20	Low-fucose	Improve FcγRIIIa binding/improve ADCC	Chronic lymphocytic leukemia	2014	2013
Vedolizumab	Entyvio	α4β7 integrin	L235A, G237	Remove Fc effector functions	Ulcerative colitis, Crohn disease	2014	2014
Nivolumab	Opdivo	PD1	S228P	Hinge stabilizing	Melanoma, non-small cell lung cancer	2015	2014
Pembrolizumab	Keytruda	PD1	S228P	Hinge stabilizing	Melanoma	2015	2014
Ixekizumab	Taltz	IL-17a	S228P	Hinge stabilizing	Psoriasis	2016	2016
Atezolizumab	Tecentriq	PD-L1	Aglycosylated (N297A)	Remove Fc effector functions	Bladder cancer	2017	2016
Dupilumab	Dupixent	IL-4Rα	S228P	Hinge stabilizing	Atopic dermatitis	2017	2017
Inotuzumab ozogamicin	BESPONSA	CD22	S228P	Hinge stabilizing	Acute lymphoblastic leukemia	2017	2017
Emicizumab	Hemlibra	Factor Ixa, X	S228P; K196Q; F296Y; E356K; K439E; L445P; G446 > del; K447 > del	Hinge stabilizing, asymmetric electrostatic steering mutations for bispecific creation	Hemophilia A	2018	2017

Benralizumab	Fasenra	IL-5Rα	Afucosylated	Improve FcγRIIIa binding/improve ADCC	Asthma	2018	2017
Durvalumab	IMFINZI	PD-L1	L234F; L235E; P331S	Remove Fc effector functions	Bladder cancer	2018	2017
Gemtuzumab ozogamicin	Mylotarg	CD33	S228P	Hinge stabilizing	Acute myeloid leukemia	2018	2017; 2000[a]
Galcanezumab (galcanezumab-gnlm)	Emgality	CGRP	S228P; F234A; L235A	Remove Fc effector functions	Migraine prevention	2018	2018
Mogamulizumab (mogamulizumab-kpkc)	Poteligeo	CCR4	Afucosylated	Improve FcγRIIIa binding/improve ADCC	Mycosis fungoides or Sézary syndrome	2018	2018
Fremanezumab (fremanezumab-vfrm)	Ajovy	CGRP	A330S; P331S	Remove Fc effector functions	Migraine prevention	2019	2018
Ravulizumab (ravulizumab-cwvz)	Ultomiris	C5	M428L; N434S	Increase half-life	Paroxysmal nocturnal hemoglobinuria	2019	2018
Cemiplimab (cemiplimab-rwlc)	Libtayo	PD-1	S228P	Hinge stabilizing	Cutaneous squamous cell carcinoma	2019	2018
Risankizumab (risankizumab-rzaa)	Skyrizi	IL-23 p19	L234A; L235A	Remove Fc effector functions	Plaque psoriasis	2019	2019

Continued

Table 5 Therapeutic antibodies with modified Fc functionality approved by FDA and EMA.—cont'd

Name	Brand name	Target	Fc modifications	Reason for Fc modifications	Indication first approved or reviewed	EMA	FDA
Eptinezumab (eptinezumab-jjmr)	VYEPTI	CGRP	Aglycosylated (N297A)	Remove Fc effector functions	Migraine prevention	EU review	2020
Inebilizumab (inebilizumab-cdon)	Uplizna	CD19	Afucosylated	Improve FcγRIIIa binding/improve ADCC	Neuromyelitis optica spectrum disorders	EU review	2020
Tafasitamab (tafasitamab-cxix)	Monjuvi, Minjuvi	CD19	S239D; I332E	Improve FcγRIIIa binding/improve ADCC	Diffuse large B-cell lymphoma	2021	2020
Belantamab mafodotin (belantamab mafodotin-blmf)	BLENREP	B-cell maturation antigen	Afucosylated	Improve FcγRIIIa binding/improve ADCC	Multiple myeloma	2020	2020
Margetuximab-cmkb	MARGENZA	HER2	F243L; R292P; Y300L; V305I; P396L	Improve FcγRIIIa binding/improve ADCC and decrease affinity for human inhibitory FcγRIIB (CD32B)	HER2+ metastatic breast cancer	NA	2020
Evinacumab	Evkeeza	Angiopoietin-like 3	S228P	Hinge stabilizing	Homozygous familial hypercholesterolemia	2021	2021
Dostarlimab	Jemerli	PD-1	S228P	Hinge stabilizing	Endometrial cancer	2021	2021

Amivantamab	RYBREVANT	EGFR, cMET	Low-fucose; K409R; F405L	Improve FcγRIIIa binding/improve ADCC; Controlled Fab arm exchange for bispecific generation	NSCLC w/EGFR exon 20 insertion mutations	EU review	2021
Teplizumab	(Pending)	CD3	L234A; L235A	Remove Fc effector functions	Type 1 diabetes	NA	In review (2nd cycle)
Narsoplimab	(Pending)	MASP-2	S228P	Hinge stabilizing	Hematopoietic stem cell transplant-associated thrombotic microangiopathies	NA	In review
Retifanlimab	(Pending)	PD-1	S228P	Hinge stabilizing	Carcinoma of the anal canal	EU review	In review (2nd cycle)
Anifrolumab, anifrolumab-fnia	Saphnelo	IFNAR1	L234F; L235E; P331S	Remove Fc effector functions	Systemic lupus erythematosus	EU review	2021
Sutimlimab	(Pending)	C1s	S228P; L235E	Hinge stabilizing; remove Fc effector functions	Cold agglutinin disease	NA	In review (2nd cycle)
Ublituximab	(Pending)	CD20	Low-fucose	Improve FcγRIIIa binding/improve ADCC	Chronic lymphocytic leukemia	NA	In review
Toripalimab	Tuoyi	PD-1	S228P	Hinge stabilizing	Nasopharyngeal carcinoma	NA	In review

Continued

Table 5 Therapeutic antibodies with modified Fc functionality approved by FDA and EMA.—cont'd

Name	Brand name	Target	Fc modifications	Reason for Fc modifications	Indication first approved or reviewed	EMA	FDA
Sintilimab	Tyvyt	PD-1	S228P	Hinge stabilizing	Non-small cell lung cancer	NA	In review
Faricimab	(Pending)	VEGF-A, Ang-2	L234A; L235A; P329G; I253A; H310A; H435A; S354C + T366W (in heavy chain A); Y349C + T366S + L368A + Y407V (in heavy chain B)	Remove Fc effector functions; Reduce half-life; CrossMab "knob-in-hole" mutations for bispecific generation	Neovascular age-related macular degeneration, diabetic macular edema	EU review	In review
Tislelizumab	(Pending)	PD-1	S228P; E233P, F234V, L235A, D265A	Hinge stabilizing; Abrogate FcγRs binding	Esophageal squamous cell carcinoma	NA	In review
Relatlimab	(Pending)	LAG-3	S228P	Hinge stabilizing	Melanoma	NA	In review
Spesolimab	(Pending)	IL-36 receptor	L234A/L235A	Remove Fc effector functions	Generalized pustular psoriasis	EU review	NA

[a]Withdrawn or marketing discontinued.
NA, not approved or in review in the EU; not approved or information on review status not available in US.
Source: The Antibody Society (www.antibodysociety.org/antibody-therapeutics-product-data). Accessed December 2021.

7. Antibody–drug conjugates

Antibody–drug conjugates are a class of biopharmaceuticals in which small molecule drugs or radioactive agents are conjugated to antibodies for targeted drug delivery; these modifications hold the promise to increase the efficacy of the small molecule/radioactive agent and reducing adverse effects based on the targeted application as well as the reduced amounts necessary (Beck et al., 2017; Larson et al., 2015; Leung et al., 2020). So far, 11 antibody–small drug conjugates, and 3 radioactive agent conjugates, have been approved for clinical use in the field of oncology for delivering cytotoxic payloads to target cells (Table 6).

Beyond oncology, antibody–drug conjugates offer a great therapeutic potential based on delivering small molecule drugs to specific cells, which is just beginning to be explored for noncancer clinical indications, including infections (Lehar et al., 2015; Mariathasan and Tan, 2017), inflammation (Everts et al., 2002; Thomsen et al., 2016; Yu et al., 2016), immunosuppression (Yu et al., 2016), vascular disorders (Kvirkvelia et al., 2015), and metabolic conditions (Lim et al., 2015). For instance, an anti-*Staphylococcus aureus* antibody conjugated to an antibiotic that is only activated once released by the phagolysosome has been demonstrated to be able to clear latent bacteria reservoirs (Lehar et al., 2015).

Although antibody conjugate technology is a very active area in antibody engineering, with more than 300 antibody conjugates in clinical trials, many aspects must still be resolved in order to further maximize its therapeutic potential. During the process of generating these therapeutic agents, it is important to pay attention to the targeting antibody, the cytotoxic payload, and the linker that holds these two components together. For instance, in targeting tumors, the targeting antibody must bind to a target antigen found either exclusively on cancer cells or overexpressed on the tumor cell as compared to healthy tissues. In addition, the antibody–drug complex must be internalized rapidly after binding its target, allowing intracellular release of the toxic drugs that kill the targeted cells (Pettinato, 2021). The linker between the molecule payload and the antibody is a very important part of the complex, as it can modulate the physiochemical properties and should minimized premature drug release yet promote selective release to the target cells (Tsuchikama and An, 2018).

Another area of great interest is the use of antibody–oligonucleotide conjugates to specifically deliver synthetic therapeutic oligonucleotides,

Table 6 Antibody-drug and antibody-radioactive agent conjugates approved by FDA and/or EMA.

Name	Brand name	Target	Payload	Payload mechanism of action	Indication first approved or reviewed	EMA	FDA
Tositumomab-I131	Bexxar	CD20	Iodine-131	I-131 produces beta and gamma emissions	Non-Hodgkin Lymphoma	NA	2003[a]
Ibritumomab tiuxetan	Zevalin	CD20	Yttrium-90	Y-90 produces beta emissions	Non-Hodgkin lymphoma	2004	2002
Certolizumab pegol	Cimzia	TNF	PEG	Extend half life	Crohn disease	2009	2008
Brentuximab vedotin	Adcetris	CD30	Monometyl auristan E (MMAE)	Tubulin polymerization inhibitor	Hodgkin lymphoma, systemic anaplastic large cell lymphoma	2012	2011
Ado-trastuzumab emtansine	Kadcyla	HER2	DM1	Tubulin polymerization inhibitor	Breast cancer	2013	2012
Inotuzumab ozogamicin	BESPONSA	CD22	Calicheamicin	DNA-binding and double breaks induction	Acute lymphoblastic leukemia	2017	2017
Gemtuzumab ozogamicin	Mylotarg	CD33	Calicheamicin	DNA-binding and double breaks induction	Acute myeloid leukemia	2018	2017; 2000[a]
Moxetumomab pasudotox (moxetumomab pasudotox-tdfk)	Lumoxiti	CD22	PE38	Immunotoxin	Hairy cell leukemia	2021	2018

Polatuzumab vedotin (polatuzumab vedotin-piiq)	Polivy	CD79b	Monomethyl Auristatin E (MMAE)	Tubulin polymerization inhibitor	Diffuse large B-cell lymphoma	2020	2019
Enfortumab vedotin (enfortumab vedotin-ejfv)	Padcev	Nectin-4	Monomethyl Auristatin E (MMAE)	Tubulin polymerization inhibitor	Urothelial cancer	EU review	2019
[fam-]trastuzumab deruxtecan, (fam-trastuzumab deruxtecan-nxki)	Enhertu	HER2	DXd	Topoisomerase I inhibitor	HER2+ metastatic breast cancer	2021	2019
Sacituzumab govitecan (sacituzumab govitecan-hziy)	TRODELVY	TROP-2	SN38	Topoisomerase I inhibitor	Triple-neg. breast cancer	2021	2020
Belantamab mafodotin (belantamab mafodotin-blmf)	BLENREP	B-cell maturation antigen	Monomethyl Auristatin F (MMAF)	Tubulin polymerization inhibitor	Multiple myeloma	2020	2020
Loncastuximab tesirine	Zynlonta	CD19	Pyrrolobenzodiazepines (PBD) dimer SG3199	DNA minor-groove binding	Diffuse large B-cell lymphoma	NA	2021
Oportuzumab monatox	(Pending)	EpCAM	PE38	Immunotoxin	Bladder cancer	NA	In review (2nd cycle)

Continued

Table 6 Antibody-drug and antibody-radioactive agent conjugates approved by FDA and/or EMA.—cont'd

Name	Brand name	Target	Payload	Payload mechanism of action	Indication first approved or reviewed	EMA	FDA
Tisotumab vedotin, tisotumab vedotin-tftv	TIVDAK	Tissue factor	Monomethyl Auristatin E (MMAE)	Tubulin polymerization inhibitor	Cervical cancer	NA	2021
Omburtamab	(Pending)	B7-H3 (CD276)	Iodine-131	I-131 produces beta and gamma emissions	Central nervous system/ leptomeningeal metastasis from neuroblastoma	EU review	NA

[a]Withdrawn or marketing discontinued; NA, not approved or in review in the EU; not approved or information on review status not available in US.
Source: The Antibody Society (www.antibodysociety.org/antibody-therapeutics-product-data). Accessed December 2021.

including antisense oligonucleotides and short interfering RNA (siRNA), to target cells (Chen et al., 2021; Juliano, 2016; Yarian et al., 2019). Although no clinical trials with antibody–oligonucleotide conjugates have been initiated to date, the progress being made in both oligonucleotide chemistry as well as in conjugate engineering will undoubtedly contribute to the development of new drugs based on these approaches (Ando et al., 2021; Benizri et al., 2019).

8. Engineered antibody fusion proteins

Another very important advance in antibody engineering has been to create antibody fusion proteins by linking different protein or peptide payloads to diverse antibody components, including to full-length antibody molecules, Fc domains, scFv fragments, and sdAbs (Silver et al., 2021). There are several advantages of generating these fusion proteins, such as increasing the half-life of pharmacologically-relevant proteins, reducing adverse events, and allowing them to be specifically delivered to the target tissue (Silver et al., 2021). The greatest therapeutic advances with this approach have been achieved using fusions of proteins to the Fc region. For instance, etanercept was the first drug fusing tumor necrosis factor receptor 2 (TNFR2) to an Fc domain to be approved by the FDA in 1998 for the treatment of rheumatoid arthritis (Goldenberg, 1999). Since then, more than a dozen Fc-fusion proteins have already been approved as therapeutics (Duivelshof et al., 2021).

A very active area of development in this field is the fusion of cytokines with antibodies (immunocytokines), which are proving useful in a number of therapeutic areas, including cancer, inflammation, and autoimmune diseases (Runbeck et al., 2021; Silver et al., 2021). Payload immunocytokines in pre-clinical or at different stages of clinical development in oncology include IL-2, IL-4, IL-15, IL-21, IFNα, and IFNβ, all of which have been fused with antibodies that target the tumor microenvironment (Corbellari et al., 2021; Villa et al., 2008). Moreover, immunocytokines have demonstrated synergy with chemotherapy and check point inhibitor therapies (Corbellari et al., 2020; Puca et al., 2020; Weide et al., 2019). The cytokine–antibody fusion proteins for inflammatory and autoimmune diseases include IL-4, IL-9, IL-10, and IFNα, although the studies for these are less developed than those for cancer (Gouyou et al., 2021; Schmid and Neri, 2019).

Other fusion proteins with antibodies at different stages of clinical development include toxins, enzymes, and neuroprotective proteins (Silver et al., 2021). Fusions of toxins to antibodies, in different formats (e.g., immunotoxins), are being developed for therapeutic use in cancer (Dashtiahangar et al., 2021) and infectious diseases (Spiess et al., 2016). The main toxins being used are modified forms of the *Pseudomonas aeruginosa* exotoxin A, diphtheria toxin, Shiga toxin, granzyme B, and granulysin (Silver et al., 2021). Targeted enzyme-replacement therapy represents another developing application of antibody engineering, which fuses an enzyme to an antibody in order to deliver the enzyme to the cytoplasm, nucleus, and/or other cellular organelles (Zhou et al., 2019). In the case of neuroprotective payloads, such as for glial cell line-derived neurotrophic factor (GDNF) and erythropoietin (EPO), preclinical studies have been developed in different pathologies (Boado et al., 2008, 2010).

Finally, it is also important to highlight the case of bintrafusp alfa, an antibody fusion protein that targets the soluble factor trap. Bintrafusp alfa consists of the extracellular domain of the TGF-βRII receptor, and it sequesters the immunosuppressive TGF-β, fused to a human IgG1 monoclonal antibody blocking PD-L1 (Gulley et al., 2021). This compound is being intensively studied for different types of cancer (Gulley et al., 2021).

9. CAR-T and CAR-NK cells

Antibody engineering is also essential in the development of CAR-T and CAR-NK cell therapies. CARs contain four main elements: (i) the extracellular region comprising an antibody, normally in an scFv format, that confers its specific antigen-binding capacity; (ii) a hinge region that provides flexibility; (iii) a transmembrane domain that anchors the CAR to the cell membrane; and (iv) one or more intracellular signaling domains (Rafiq et al., 2020; Sterner and Sterner, 2021). Modifications in these intracellular domains have been a central aspect for the progress of CAR technology. First-generation CARs had only the intracellular signaling domain CD3ζ (Eshhar et al., 1993), while the second generation of CARs added a costimulatory domain (e.g., CD28, 4-1BB) in series with CD3ζ (Maher et al., 2002). The incorporation of a second costimulation domain in series with CD3ζ has given rise to third-generation CARs (Pulè et al., 2005). The fourth generation CARs include immune modifiers that can shape the tumor microenvironment (Chmielewski and Abken, 2015). In addition to CAR components that are common to CAR-T cells and CAR-NK cells, using other activation signaling domains (such as DAP10, and DAP12), and NKG2D as a receptor is also being explored for CAR-NK cells (Lu et al., 2021) (Fig. 6).

Beyond conferring antigen specificity, some characteristics of the scFv will affect CAR functions. For instance, it is important to have enough scFv affinity for the target antigen, as it much induce CAR signaling and cell activation, yet it is essential to avoid excessive affinity, which could trigger activation-induced cell death of CAR-expressing cells (Hudecek et al., 2013; Liu et al., 2015). Other aspects to take into consideration when selecting an scFv for CAR construction are (i) the density of the target antigen, (ii) the location of the recognized epitope (Smith et al., 2018), and (iii) avoiding scFv that transmit tonic signaling independent of their interactions with the antigen (Ajina and Maher, 2018).

CAR-focused therapies have shown great efficacy in numerous hematological tumors by using CAR-T directed at CD19. Accordingly, four CAR-T cells targeting CD19 have been approved by the FDA to treat some hematological malignancies: tisagenlecleucel (Kymriah), axicabtagén ciloleucel (Yescarta), brexucabtagene autoleucel (Tecartus), and lisocabtagene

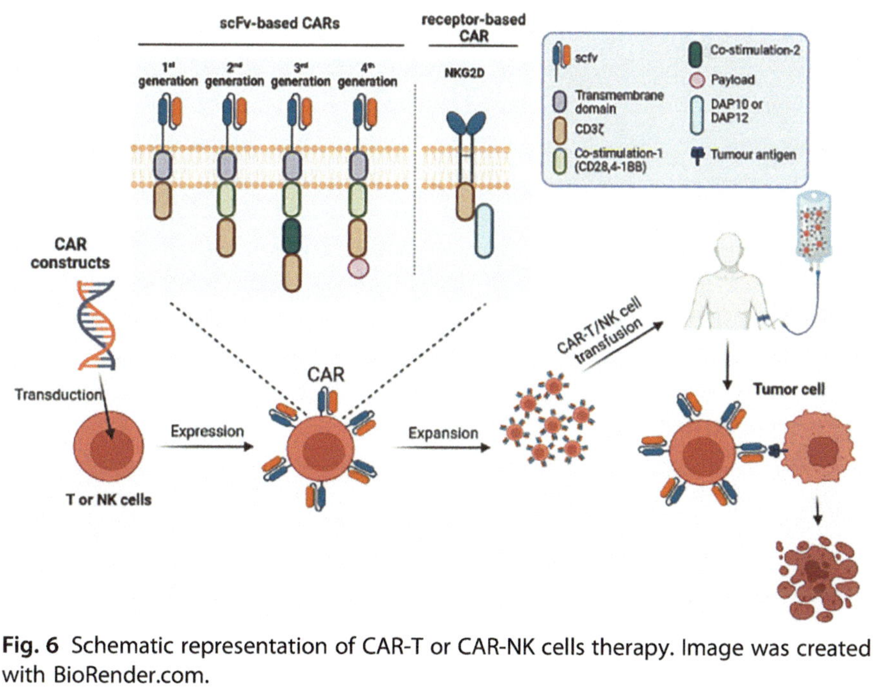

Fig. 6 Schematic representation of CAR-T or CAR-NK cells therapy. Image was created with BioRender.com.

maraleucel (Breyanzi). A major effort is now being made to improve CAR technology and to translate the hematological malignancy successes to solid tumors as well. Currently, there are 726 clinical trials registered with CAR-T cells, and 17 clinical trials with CAR-NK cells (https://clinicaltrials.gov/). Although more clinical trials are underway with CAR-T cells than with CAR-NK cells, CAR-NK cells can offer several advantages over CAR-T, such as have fewer adverse effects and enhancing cytotoxic killing capacity (Lu et al., 2021).

10. Conclusions

Since the hybridoma technology was initiated in 1975 by Köhler and Milstein, we have witnessed a great development of antibody engineering, from the early stages of antibody humanization to the development of antibody-based molecules in very sophisticated formats. These new formats aim to increase therapeutic efficacy by modifying antibody sizes, increasing the number of recognized antigens, modifying the Fc region, delivering small-molecule payloads to specific cell types, fusing other proteins to antibodies, and/or forming chimeric antigen receptors. These technological breakthroughs, together with the exquisite specificity and affinity of the antibodies for their targets, have led to a paradigm shift in the pharmaceutical industry—from the development of mostly chemical-based compounds, to the development of antibody-based therapies. In fact, antibody-based therapies are twice as likely to succeed in clinical trials as small-chemical compounds. Today, antibody-based therapies play a key role in the treatment of various diseases, including cancer, autoimmune and neurological disorders, and infectious and metabolic diseases, with some antibodies reaching the top 10 bestselling pharmaceutical drugs in some years; these successes are expected to increase as ongoing clinical trials are completed and as further improvements are made in engineered antibody efficacy.

Acknowledgments

The Yélamos lab is funded by the Spanish Ministerio de Ciencia e Innovación (grant PID2020-112526RB-I00; funded by MCIN/AEI/10.13039/501100011033).

References

Ajina, A., Maher, J., 2018. Strategies to address chimeric antigen receptor tonic signaling. Mol. Cancer Ther. 17, 1795–1815.
Ando, Y., et al., 2021. Enzymatic ligation of an antibody and arginine 9 peptide for efficient and cell-specific siRNA delivery. Sci. Rep. 11, 21882.

Arbabi Ghahroudi, M., et al., 1997. Selection and identification of single domain antibody fragments from camel heavy-chain antibodies. FEBS Lett. 414, 521–526.
Balhuizen, A., et al., 2017. A nanobody-based tracer targeting DPP6 for non-invasive imaging of human pancreatic endocrine cells. Sci. Rep. 7, 15130.
Barnhart, B.C., Quigley, M., 2017. Role of fc-FcγR interactions in the antitumor activity of therapeutic antibodies. Immunol. Cell Biol. 95, 340–346.
Beck, A., et al., 2017. Strategies and challenges for the next generation of antibody-drug conjugates. Nat. Rev. Drug Discov. 16, 315–337.
Benizri, S., et al., 2019. Bioconjugated oligonucleotides: recent developments and therapeutic applications. Bioconjug. Chem. 30, 366–383.
Blanco, B., Domínguez-Alonso, C., Alvarez-Vallina, L., 2021. Bispecific immunomodulatory antibodies for cancer immunotherapy. Clin. Cancer Res. 27, 5457–5464.
Boado, R.J., et al., 2008. GDNF fusion protein for targeted-drug delivery across the human blood-brain barrier. Biotechnol. Bioeng. 100, 387–396.
Boado, R.J., et al., 2010. Drug targeting of erythropoietin across the primate blood-brain barrier with an IgG molecular trojan horse. J. Pharmacol. Exp. Ther. 333, 961–969.
Borrok, M.J., et al., 2017. An 'fc-silenced' IgG1 format with extended half-life designed for improved stability. J. Pharm. Sci. 106, 1008–1017.
Boulianne, G.L., Hozumi, N., Shulman, M.J., 1984. Production of functional chimaeric mouse/human antibody. Nature 312, 643–646.
Brezski, R.J., Georgiou, G., 2016. Immunoglobulin isotype knowledge and application to fc engineering. Curr. Opin. Immunol. 40, 62–69.
Brinkmann, U., Kontermann, R.E., 2017. The making of bispecific antibodies. MAbs 9, 182–212.
Chanier, T., Chames, P., 2019. Nanobody engineering: toward next generation immunotherapies and immunoimaging of cancer. Antibodies (Basel, Switzerland) 8, 13.
Chen, J., et al., 2019. A BAP31 intrabody induces gastric cancer cell death by inhibiting p27 kip1 proteasome degradation. Int. J. Cancer 144, 2051–2062.
Chen, S., et al., 2021. Anti-EpCAM functionalized graphene oxide vector for tumor targeted siRNA delivery and cancer therapy. Asian J. Pharm. Sci. 16, 598–611.
Chiu, M.L., et al., 2019. Antibody structure and function: the basis for engineering therapeutics. Antibodies (Basel, Switzerland) 8, 55.
Chmielewski, M., Abken, H., 2015. TRUCKs: the fourth generation of CARs. Expert Opin. Biol. Ther. 15, 1145–1154.
Corbellari, R., et al., 2020. The immunocytokine L19-TNF eradicates sarcomas in combination with chemotherapy agents or with immune check-point inhibitors. Anticancer Drugs 31, 799–805.
Corbellari, R., et al., 2021. A novel antibody–IL15 fusion protein selectively localizes to tumors, synergizes with TNF-based Immunocytokine, and inhibits metastasis. Mol. Cancer Ther. 20, 859–871.
Corti, D., Lanzavecchia, A., 2013. Broadly neutralizing antiviral antibodies. Annu. Rev. Immunol. 31, 705–742.
Dashtiahangar, M., et al., 2021. Anti-cancer immunotoxins, challenges, and approaches. Curr. Pharm. Des. 27, 932–941.
Den Broeder, A., et al., 2002. A single dose, placebo controlled study of the fully human anti-tumor necrosis factor-alpha antibody adalimumab (D2E7) in patients with rheumatoid arthritis. J. Rheumatol. 29, 2288–2298.
Duivelshof, B.L., et al., 2021. Therapeutic fc-fusion proteins: current analytical strategies. J. Sep. Sci. 44, 35–62.
Emmons, C., Hunsicker, L.G., 1987. Muromonab-CD3 (Orthoclone OKT3): the first monoclonal antibody approved for therapeutic use. Iowa Med. 77, 78–82.

Eshhar, Z., et al., 1993. Specific activation and targeting of cytotoxic lymphocytes through chimeric single chains consisting of antibody-binding domains and the gamma or zeta subunits of the immunoglobulin and T-cell receptors. Proc. Natl. Acad. Sci. U. S. A. 90, 720–724.

Everts, M., et al., 2002. Selective intracellular delivery of dexamethasone into activated endothelial cells using an E-selectin-directed immunoconjugate. J. Immunol. 168, 883–889.

Gibson, T.B., Ranganathan, A., Grothey, A., 2006. Randomized phase III trial results of panitumumab, a fully human anti-epidermal growth factor receptor monoclonal antibody, in metastatic colorectal cancer. Clin. Colorectal Cancer 6, 29–31.

Golay, J., et al., 2013. Glycoengineered CD20 antibody obinutuzumab activates neutrophils and mediates phagocytosis through CD16B more efficiently than rituximab. Blood 122, 3482–3491.

Goldenberg, M.M., 1999. Etanercept, a novel drug for the treatment of patients with severe, active rheumatoid arthritis. Clin. Ther. 21, 75–87.

Gouyou, B., et al., 2021. Therapeutic evaluation of antibody-based targeted delivery of interleukin 9 in experimental pulmonary hypertension. Int. J. Mol. Sci. 22, 3460.

Gulley, J.L., et al., 2021. Dual Inhibition of TGF-β and PD-L1: A Novel Approach to Cancer Treatment. Mol, Oncol.

Heiss, M.M., et al., 2010. The trifunctional antibody catumaxomab for the treatment of malignant ascites due to epithelial cancer: results of a prospective randomized phase II/III trial. Int. J. Cancer 127, 2209–2221.

Hozumi, N., Tonegawa, S., 1976. Evidence for somatic rearrangement of immunoglobulin genes coding for variable and constant regions. Proc. Natl. Acad. Sci. U. S. A. 73, 3628–3632.

Hudecek, M., et al., 2013. Receptor affinity and extracellular domain modifications affect tumor recognition by ROR1-specific chimeric antigen receptor T cells. Clin. Cancer Res. 19, 3153–3164.

Jones, P.T., et al., 1986. Replacing the complementarity-determining regions in a human antibody with those from a mouse. Nature 321, 522–525.

Juliano, R.L., 2016. The delivery of therapeutic oligonucleotides. Nucleic Acids Res. 44, 6518–6548.

Kasono, K., et al., 1998. Functional knock-out of c-myb by an intracellular anti-c-Myb single-chain antibody. Biochem. Biophys. Res. Commun. 251, 124–130.

Kempeni, J., 1999. Preliminary results of early clinical trials with the fully human anti-TNFalpha monoclonal antibody D2E7. Ann. Rheum. Dis. 58 (Suppl), 1.

Köhler, G., Milstein, C., 1975. Continuous cultures of fused cells secreting antibody of predefined specificity. Nature 256, 495–497.

Kvirkvelia, N., et al., 2015. Human anti-α3(IV)NC1 antibody drug conjugates target glomeruli to resolve nephritis. Am. J. Physiol. Renal Physiol. 309, F680–F684.

Kwakkenbos, M.J., van Helden, P.M., Beaumont, T., Spits, H., 2016. Stable long-term cultures of self-renewing B cells and their applications. Immunol. Rev. 270, 65–77.

Labrijn, A.F., Aalberse, R.C., Schuurman, J., 2008. When binding is enough: nonactivating antibody formats. Curr. Opin. Immunol. 20, 479–485.

Labrijn, A.F., et al., 2019. Bispecific antibodies: a mechanistic review of the pipeline. Nat. Rev. Drug Discov. 18, 585–608.

Larson, S.M., et al., 2015. Radioimmunotherapy of human tumours. Nat. Rev. Cancer 15, 347–360.

Lazar, G.A., et al., 2006. Engineered antibody fc variants with enhanced effector function. Proc. Natl. Acad. Sci. U. S. A. 103, 4005–4010.

Lehar, S.M., et al., 2015. Novel antibody-antibiotic conjugate eliminates intracellular S. aureus. Nature 527, 323–328.

Leung, D., et al., 2020. Antibody conjugates-recent advances and future innovations. Antibodies (Basel, Switzerland) 9, 2.
Lim, R.K.V., et al., 2015. Targeted delivery of LXR agonist using a site-specific antibody-drug conjugate. Bioconjug. Chem. 26, 2216–2222.
Liu, X., et al., 2015. Affinity-tuned ErbB2 or EGFR chimeric antigen receptor T cells exhibit an increased therapeutic index against tumors in mice. Cancer Res. 75, 3596–3607.
Lo, M., et al., 2017. Effector-attenuating substitutions that maintain antibody stability and reduce toxicity in mice. J. Biol. Chem. 292, 3900–3908.
Lonberg, N., 2005. Human antibodies from transgenic animals. Nat. Biotechnol. 23, 1117–1125.
Lu, R.M., et al., 2020. Development of therapeutic antibodies for the treatment of diseases. J. Biomed. Sci. 27, 1–30.
Lu, H., Zhao, X., Li, Z., Hu, Y., Wang, H., 2021. From CAR-T cells to CAR-NK cells: a developing immunotherapy method for hematological malignancies. Front. Oncol. 11, 720501.
Ma, J., et al., 2021. Bispecific antibodies: from research to clinical application. Front. Immunol. 12, 626616.
Maher, J., et al., 2002. Human T-lymphocyte cytotoxicity and proliferation directed by a single chimeric TCRzeta /CD28 receptor. Nat. Biotechnol. 20, 70–75.
Mariathasan, S., Tan, M.W., 2017. Antibody-antibiotic conjugates: a novel therapeutic platform against bacterial infections. Trends Mol. Med. 23, 135–149.
Marschall, A.L.J., et al., 2011. Targeting antibodies to the cytoplasm. MAbs 3, 3–16.
Mendez, M.J., et al., 1997. Functional transplant of megabase human immunoglobulin loci recapitulates human antibody response in mice. Nat. Genet. 15, 146–156.
Merchant, A.M., et al., 1998. An efficient route to human bispecific IgG. Nat. Biotechnol. 16, 677–681.
Moroni, M., et al., 2005. Gene copy number for epidermal growth factor receptor (EGFR) and clinical response to antiEGFR treatment in colorectal cancer: a cohort study. Lancet Oncol. 6, 279–286.
Morrison, S.L., et al., 1984. Chimeric human antibody molecules: mouse antigen-binding domains with human constant region domains. Proc. Natl. Acad. Sci. U. S. A. 81, 6851–6855.
Mullard, A., 2021. FDA approves 100th monoclonal antibody product. Nat. Rev. Drug Discov. 20, 491–495.
Munro, S., Pelham, H.R.B., 1987. A C-terminal signal prevents secretion of luminal ER proteins. Cell 48, 899–907.
Muyldermans, S., 2021. Applications of nanobodies. Annu. Rev. Anim. Biosci. 9, 401–421.
Neuberger, M.S., et al., 1998. Monitoring and interpreting the intrinsic features of somatic hypermutation. Immunol. Rev. 162, 107–116.
Newnham, L.E., et al., 2015. Functional inhibition of β-catenin-mediated Wnt signaling by intracellular VHH antibodies. MAbs 7, 180–191.
Nisonoff, A., Rivers, M.M., 1961. Recombination of a mixture of univalent antibody fragments of different specificity. Arch. Biochem. Biophys. 93, 460–462.
Page, D.B., et al., 2014. Immune modulation in cancer with antibodies. Annu. Rev. Med. 65, 185–202.
Pettinato, M.C., 2021. Introduction to antibody-drug conjugates. Antibodies (Basel, Switzerland) 10, 42.
Puca, E., et al., 2020. The antibody-based delivery of interleukin-12 to solid tumors boosts NK and CD8+ T cell activity and synergizes with immune checkpoint inhibitors. Int. J. Cancer 146, 2518–2530.
Pulè, M.A., et al., 2005. A chimeric T cell antigen receptor that augments cytokine release and supports clonal expansion of primary human T cells. Mol. Ther. 12, 933–941.

Quevedo, C.E., et al., 2018. Small molecule inhibitors of RAS-effector protein interactions derived using an intracellular antibody fragment. Nat. Commun. 9, 3169.

Rafiq, S., Hackett, C.S., Brentjens, R.J., 2020. Engineering strategies to overcome the current roadblocks in CAR T cell therapy. Nat. Rev. Clin. Oncol. 17, 147–167.

Redchuk, T.A., et al., 2020. Optogenetic regulation of endogenous proteins. Nat. Commun. 11, 605.

Ridgway, J.B., Presta, L.G., Carter, P., 1996. 'Knobs-into-holes' engineering of antibody CH3 domains for heavy chain heterodimerization. Protein Eng. 9, 617–621.

Runbeck, E., et al., 2021. Utilizing Immunocytokines for cancer therapy. Antibodies (Basel, Switzerland) 10, 1–23.

Schmid, A.S., Neri, D., 2019. Advances in antibody engineering for rheumatic diseases. Nat. Rev. Rheumatol. 15, 197–207.

Schuurman, J., et al., 1999. Normal human immunoglobulin G4 is bispecific: it has two different antigen-combining sites. Immunology 97, 693–698.

Scully, M., et al., 2019. Caplacizumab treatment for acquired thrombotic thrombocytopenic purpura. N. Engl. J. Med. 2019 (380), 335–346.

Shields, R.L., et al., 2001. High resolution mapping of the binding site on human IgG1 for fc gamma RI, fc gamma RII, fc gamma RIII, and FcRn and design of IgG1 variants with improved binding to the fc gamma R. J. Biol. Chem. 276, 6591–6604.

Shields, R.L., et al., 2002. Lack of fucose on human IgG1 N-linked oligosaccharide improves binding to human Fcgamma RIII and antibody-dependent cellular toxicity. J. Biol. Chem. 277, 26733–26740.

Silver, A.B., et al., 2021. Engineered antibody fusion proteins for targeted disease therapy. Trends Pharmacol. Sci. 42, 1064–1081.

Simoons, M.L., et al., 1994. Randomized trial of a GPIIb/IIIa platelet receptor blocker in refractory unstable angina. European Cooperative Study Group. Circulation 89, 596–603.

Smith, S.L., 1996. Ten years of Orthoclone OKT3 (muromonab-CD3): a review. J. Transpl. Coord. 6, 109–119.

Smith, E.L., et al., 2018. Development and evaluation of an optimal human single-chain variable fragment-derived BCMA-targeted CAR T cell vector. Mol. Ther. 26, 1447–1456.

Sondermann, P., Szymkowski, D.E., 2016. Harnessing fc receptor biology in the design of therapeutic antibodies. Curr. Opin. Immunol. 40, 78–87.

Spiess, K., et al., 2016. The future of antiviral immunotoxins. J. Leukoc. Biol. 99, 911–925.

Steeland, S., Vandenbroucke, R.E., Libert, C., 2016. Nanobodies as therapeutics: big opportunities for small antibodies. Drug Discov. Today 21, 1076–1113.

Sterner, R.C., Sterner, R.M., 2021. CAR-T cell therapy: current limitations and potential strategies. Blood. Cancer J. 11.

Strube, R.W., Chen, S.Y., 2002. Characterization of anti-cyclin E single-chain Fv antibodies and intrabodies in breast cancer cells: enhanced intracellular stability of novel sFv-F(c) intrabodies. J. Immunol. Methods 263, 149–167.

Suzuki, M., Kato, C., Kato, A., 2015. Therapeutic antibodies: their mechanisms of action and the pathological findings they induce in toxicity studies. J. Toxicol. Pathol. 28, 133.

Tanaka, T., Williams, R.L., Rabbitts, T.H., 2007. Tumour prevention by a single antibody domain targeting the interaction of signal transduction proteins with RAS. EMBO J. 26, 3250–3259.

Thomsen, K.L., et al., 2016. Anti-CD163-dexamethasone conjugate inhibits the acute phase response to lipopolysaccharide in rats. World J. Hepatol. 8, 726–730.

Tiller, T., et al., 2008. Efficient generation of monoclonal antibodies from single human B cells by single cell RT-PCR and expression vector cloning. J. Immunol. Methods 329, 112–124.

Tsuchikama, K., An, Z., 2018. Antibody-drug conjugates: recent advances in conjugation and linker chemistries. Protein Cell 9, 33–46.
Vafa, O., et al., 2014. An engineered fc variant of an IgG eliminates all immune effector functions via structural perturbations. Methods 65, 114–126.
Van Erp, E.A., et al., 2019. Fc-mediated antibody effector functions during respiratory syncytial virus infection and disease. Front. Immunol. 10, 548.
Van Landuyt, L., et al., 2019. Customized protein glycosylation to improve biopharmaceutical function and targeting. Curr. Opin. Biotechnol. 60, 17–28.
Villa, A., et al., 2008. A high-affinity human monoclonal antibody specific to the alternatively spliced EDA domain of fibronectin efficiently targets tumor neo-vasculature in vivo. Int. J. Cancer 122, 2405–2413.
Vincenti, F., et al., 1998. Interleukin-2-receptor blockade with daclizumab to prevent acute rejection in renal transplantation. Daclizumab triple therapy study group. N. Engl. J. Med. 338, 161–165.
Vincke, C., et al., 2009. General strategy to humanize a camelid single-domain antibody and identification of a universal humanized nanobody scaffold. J. Biol. Chem. 284, 3273–3284.
Walker, M.R., et al., 1989. Aglycosylation of human IgG1 and IgG3 monoclonal antibodies can eliminate recognition by human cells expressing fc gamma RI and/or fc gamma RII receptors. Biochem. J. 259, 347–353.
Weide, B., et al., 2019. A phase II study of the L19IL2 immunocytokine in combination with dacarbazine in advanced metastatic melanoma patients. Cancer Immunol. Immunother. 68, 1547–1559.
Wheeler, Y.Y., Chen, S.Y., Sane, D.C., 2003. Intrabody and intrakine strategies for molecular therapy. Mol. Ther. 8, 355–366.
Winter, G., 2019. Harnessing evolution to make medicines (Nobel lecture). Angew. Chem. Int. Ed. Engl. 58, 14438–14445.
Winter, G., Milstein, C., 1991. Man-made antibodies. Nature 349, 293–299.
Winter, G., et al., 1994. Making antibodies by phage display technology. Annu. Rev. Immunol. 12, 433–455.
Wu, Y., et al., 2020. A cyclin D1-specific single-chain variable fragment antibody that inhibits HepG2 cell growth and proliferation. Biotechnol. J. 15.
Xie, Y.J., et al., 2019. Nanobody-based CAR T cells that target the tumor microenvironment inhibit the growth of solid tumors in immunocompetent mice. Proc. Natl. Acad. Sci. U. S. A. 116, 7624–7631.
Xie, Y.J., et al., 2020. Improved antitumor efficacy of chimeric antigen receptor T cells that secrete single-domain antibody fragments. Cancer Immunol. Res. 8, 518–529.
Yarian, F., et al., 2019. Antibody-drug therapeutic conjugates: potential of antibody-siRNAs in cancer therapy. J. Cell. Physiol. 234, 16724–16738.
Yélamos, J., et al., 1995. Targeting of non-Ig sequences in place of the V segment by somatic hypermutation. Nature 376, 225–229.
Yu, S., et al., 2016. Targeted delivery of an anti-inflammatory PDE4 inhibitor to immune cells via an antibody-drug conjugate. Mol. Ther. 24, 2078–2089.
Zhang, C., et al., 2020. Applying antibodies inside cells: principles and recent advances in neurobiology, virology and oncology. BioDrugs 34, 435–462.
Zhou, Z., et al., 2019. Antibody-mediated enzyme therapeutics and applications in glycogen storage diseases. Trends Mol. Med. 25, 1094–1109.

CHAPTER TWO

Opportunities and challenges of bi-specific antibodies

Aina Segués[a,b,†], Shuyu Huang[a,b,†], Alice Sijts[b], Pedro Berraondo[c,d,e], and Dietmar M. Zaiss[a,f,g,h,*]

[a]Institute of Immunology and Infection Research, School of Biological Sciences, University of Edinburgh, Edinburgh, United Kingdom
[b]Faculty of Veterinary Medicine, Department of Infectious Diseases and Immunology, Utrecht University, Utrecht, The Netherlands
[c]Program of Immunology and Immunotherapy, Cima Universidad de Navarra, Pamplona, Spain
[d]Centro de Investigación Biomédica en Red de Cáncer (CIBERONC), Madrid, Spain
[e]Navarra Institute for Health Research (IDISNA), Pamplona, Spain
[f]Department of Immune Medicine, University Regensburg, Regensburg, Germany
[g]Institute of Clinical Chemistry and Laboratory Medicine, University Hospital Regensburg, Regensburg, Germany
[h]Institute of Pathology, University Regensburg, Regensburg, Germany
*Corresponding author: e-mail address: dietmar.zaiss@ukr.de

Contents

† Authors contributed equally.

International Review of Cell and Molecular Biology, Volume 369
ISSN 1937-6448
https://doi.org/10.1016/bs.ircmb.2022.05.001

Abstract

The recent clinical approval of different Bi-specific antibodies (BsAbs) has revealed the great therapeutic potential of this novel class of biologicals. For example, the bispecific T-cell engager (BiTE), Blinatumomab, demonstrated the unique capacity of BsAbs to link T-cells with tumor cells, inducing targeted tumor cell removal. Additionally, Amivantamab, recognizing the EGFR and cMet *in cis*, revealed a substantial improvement of therapeutic efficacy by concomitantly targeting two tumor antigens. *Cis*-targeting BsAbs furthermore allow discerning cell populations which concurrently express two antigens, for which each antigen expression pattern in itself might not be selective. In this way, BsAbs harbor the great prospect of being more specific and showing fewer side effects than monoclonal antibodies. Nevertheless, BsAbs have also faced major obstacles, for instance, in ensuring reliable assembly and clinical-grade purification. In this review, we summarize the different available antibody platforms currently used for the generation of IgG-like and non-IgG-like BsAbs and explain which approaches have been used to assemble those BsAbs which are currently approved for clinical application. By focusing on the example of regulatory T-cells (Tregs) and the different, ongoing approaches to develop BsAbs specifically targeting Tregs within the tumor microenvironment, our review highlights the huge potential as well as the pitfalls BsAb face in order to emerge as one of the most effective therapeutic biologicals targeting desired cell populations in a highly selective way. Such BsAb may improve treatment efficacy and reduce side effects, thereby opening novel treatment opportunities for a range of different diseases, such as cancer or autoimmune diseases.

1. Introduction

The clinical application of monoclonal antibodies has transformed the treatment of a wide range of diseases. Prominent successes have been the use of TNF-inhibitors for therapy of inflammatory diseases (Zhang et al., 2021b), and so-called immune check-point inhibitors in the field of tumor immunotherapy (Chen et al., 2021). Nonetheless, the clinical success of monoclonal antibody-based therapeutic approaches has slowed in recent years. In many cases, treatment turned out to be associated with severe side-effects as the effects of injected antibodies are often not focused on the site of inflammation or the tumor itself but may mediate systemic effects. Furthermore, many attractive target antigens are expressed in a more ubiquitous way, thus their expression is not limited to for instance specific tumor cells or leukocyte subpopulations, such as regulatory T-cells. One rather appealing approach to increase the specificity of antibody-based treatment is the generation of bispecific antibodies (BsAbs), whereby one antibody is recognizing two different antigens at the same time. Already in the 1960s, the very first BsAb targeting two different antigens with one antibody was described by Nisonoff et al. (1960). At that stage, two rabbit antigen-binding fragments were combined from two different polyclonal sera through mild re-oxidation (Nisonoff et al., 1960). However, the advent of recombinant antibody design has dramatically enhanced the development of BsAbs. Since the marketing of Blinatumomab, a fragment based BsAb targeting CD19 and CD3 *in trans*, and of Emicizumab, a BsAb that mimics the function of the coagulation factor VIII, their sales have increased continuously. In 2019, Blinatumomab's sales reached $312 million and the sales of Emicizumab $1.49 billion (Hosseini et al., 2021). It is currently estimated that the BsAb market scale will surpass $8 billion by 2025. Accordingly, the growth rate of the number of BsAb in the developmental pipeline has been estimated to reach three times that of conventional monoclonal antibody drugs in the next few years.

In this review, we will discuss the progress that bispecific antibody design has made in recent years, their opportunities and, in particular, the current limitations their design is facing.

2. *Cis*- versus *trans*-targeting bispecific antibodies

Compared to monospecific antibodies (mAbs), BsAbs potentially harbor several advantages. In particular, BsAbs have the capability to target cells either in a *cis*- or in a *trans*-binding orientation. During *trans*-binding, the antibody recognizes two different antigens, each expressed on a different cell population. Thus, such BsAbs can link two different cell populations with each other. One of the most prominent examples of such *trans*-binding BsAb are the bispecific T-cell engagers (BiTE). For the construction of these cell-engaging BsAbs, a T-cell or NK-cell specific antibody fragment is linked with a tumor antigen recognizing antibody (DiLillo et al., 2021) (Fig. 1). In this way, within a tumor the cell engager cross-links tumor cells with T- or NK-cells, thereby targeting their cellular cytotoxicity toward the tumor cell and, thus, inducing the killing of the recognized tumor cell. Following the clinical success of Blinatumomab, the first BiTE approved for clinical application, cell engagers constructs have reached much attention with a wide range of different constructs targeting different tumor antigens (Tian et al., 2021).

In contrast to the *trans*-binding, cis-binding BsAbs are targeted to antigens expressed on the very same cell. By dual targeting, BsAbs could increase target selectivity by concurrent binding to two antigens. Given the affinity for each antigen were low enough for the antibody to bind only weakly to

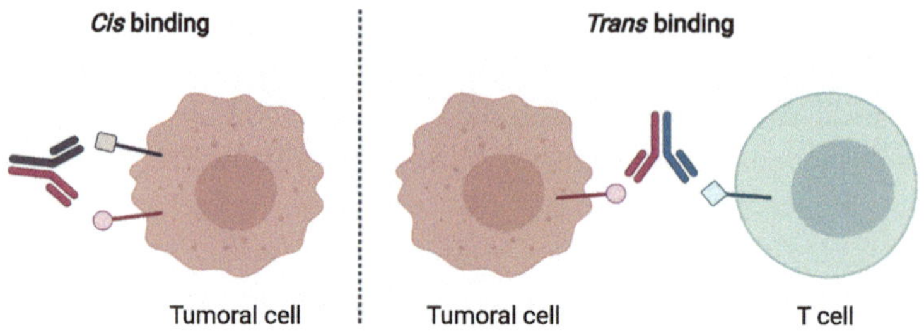

Fig. 1 Scheme illustrating Bi-specific Antibodies binding in a cis- and a trans-mode. Bi-specific antibodies recognize two antigens. If both antigens are expressed on the same cell surface, then a cis-binding antibody recognizes both antigens on the same, the cis-side. However, if both antigens are expressed on two different cells, then a trans-binding antibody can form a link in between two cells, with the second antigen being expressed on the other, the trans-side.

target cells that express only one of the antigens but strongly to cells that express both antigens, then such a BsAb could be highly selective for target cells presenting both antigens on their cell surface (Mazor et al., 2015). In this way, these cis-binding BsAbs could be particularly attractive for clinical application, as specific tumor antigens could be included in the repertoire of clinically targeted antigens, despite the fact that also other cell types may occasionally express one of the two targeted antigens. In a similar way, bispecific, depleting antibodies could then be specifically targeted toward selected leukocyte populations, such as regulatory T-cells, based on the combined expression of antigens. Such a selectivity of treatment mediated by bi-specific antibodies could then substantially improve the efficacy of treatment and diminish the side-effects currently observed by the use of several monoclonal antibodies.

3. Technical challenges in the design and production of bispecific antibodies

Although Nisonoff et al. had already demonstrated the feasibility of the concept of bispecific antibodies in the 1960s, the first market approval for a therapeutic BsAb was almost 50 years later, in 2009. This delay was due to a number of challenges that the design of such BsAbs is facing. Main challenges that emerged were the specificity of cis-targeting constructs and the manufacturing of the constructs (Nisonoff et al., 1960). When comparing the mechanism of action of monospecific with that of bispecific antibodies, a number of differences become apparent (Table 1). mAbs are

Table 1 Comparison between monospecific and bispecific antibodies.

	Monospecific antibody	Bispecific antibody
MOA	Target one antigen of interest, inhibit/stimulate one signal pathway, Fc mediated effect	Emerging MOAs only for BsAbs: bridging immune cells and tumor cells; inhibit/ stimulate multiple signal pathways simultaneously
Antigen recognition	Single epitope	Multiple epitopes
Specificity	High	Very high
Stability	Medium	Relatively low; more sensitive to physical/chemical conditions
Standardized production	Relatively easy; little batch difference	Relatively difficult; potential batch difference
Production cost	Medium	Relatively high

Abbreviations: MOA: mechanism of action.

relatively easy to produce in large scale and tend to be more stable than BsAbs. This is mainly due to the fact that the modified sequences of BsAbs, which are used to enhance combination of two different arms, are more sensitive to physical and chemical conditions. Also, the large-scale production methods for mAbs are easier compared to BsAbs and less differences are present between batches which increases the reliability of mAbs over BsAbs during clinical application (Byrne et al., 2013).

Faced with these challenges, several different technical approaches were established that would allow the reliable production of highly pure and stable BsAbs (Labrijn et al., 2019).

In 1983, the hybrid hybridoma (quadroma) technology was invented for bispecific antibody production (Milstein and Cuello, 1983). In this approach, two hybridoma cell lines, expressing a different monoclonal antibody each, were fused. In this fused hybridoma randomly formed heavy/ light chain pairs are formed during assembly. The theoretical yield of desired bispecific antibodies produced by this approach was assumed to be around an eighth of all produced antibodies. Nevertheless, during purification the different sets of BsAbs turned out to be extremely difficult to separate, and the production of pure BsAbs without the presence of non-desired by-products remained a major obstacle (Klein et al., 2012). Over the last decades, paralleled by advances in protein engineering, several different approaches and production platforms have been developed that enforce the correct association of two different antibody pairs. Thereby one has to differentiate between so-called "IgG-like" and "non-IgG-like" BsAb, whereby the IgG-like BsAbs in principle retain the classical heavy and light chain structure, while the non-IgG-like BsAbs leave that structure entirely and combine two or more antigen-binding domains, which could be antigen-binding fragment (Fab), single-chain variable fragment (scFv) or single-domain antibody (sdAb), also known as a nanobody. The IgG-like BsAbs have the substantial advantage that by interacting with the FcR expressed on different types of immune cells they are able to induce secondary immune functions, such as antibody-dependent cellular cytotoxicity (ADCC), antibody-dependent cellular phagocytosis (ADCP) and complement-dependent cytotoxicity (CDC). By enhanced engineering, such as interfering with the glycosylation of the heavy chain, the interaction of these BsAbs with specific FcRs can be altered and, consequently, the clinical efficacy of the used antibody substantially improved (Vukovic et al., 2021).

In contrast, non-IgG-like BsAbs have the edge over IgG-like ones with regard to immunogenicity and tissue-penetrating capacity (Fan et al., 2015). The T-cell engager Blinatumomab is a prominent example of an approved non-IgG-like BsAb. The bispecific T cell engager (BiTE) platform used for its construction consists of two scFv parts, one targeting CD3 and a second the cell surface-expressed tumor-associated antigen (TAA) CD19, which are linked by a G4S linker (Fig. 2A). The resulting 55 kDa BsAb, Blinatumomab showed an impressive efficacy at very low doses in patients with non-Hodgkin lymphoma (NHL) and relapsed and/or refractory (r/r) acute lymphoblastic leukemia (ALL) (Loffler et al., 2000; Rogala et al., 2015). Most interestingly, clinical trials using a similar antibody targeting CD3 and CD19 (MGD-011/Duvortuxizumab) had to be ceased due to safety concerns over neurological events occurring in some patients during treatment for ALL in phase I studies (NCT02848911) and r/r NHL (NCT02106091) (Reusch et al., 2015). This antibody was based on the dual-affinity retargeting (DART) platform, which is similar to the BiTE platform but covalently links two Fab-fragments through C-terminal cysteine residues (Fig. 2B).

A number of additional non-IgG-like antibodies are currently in clinical trial. Prominently, the company Ablynx uses its nanobody platform to construct Bi-Nanobodies. This platform connects three nanobody moieties into a trivalent molecule linked by two flexible poly-Glycine/Serine linkers. Two of these nanobody moieties specifically recognize therapeutic targets

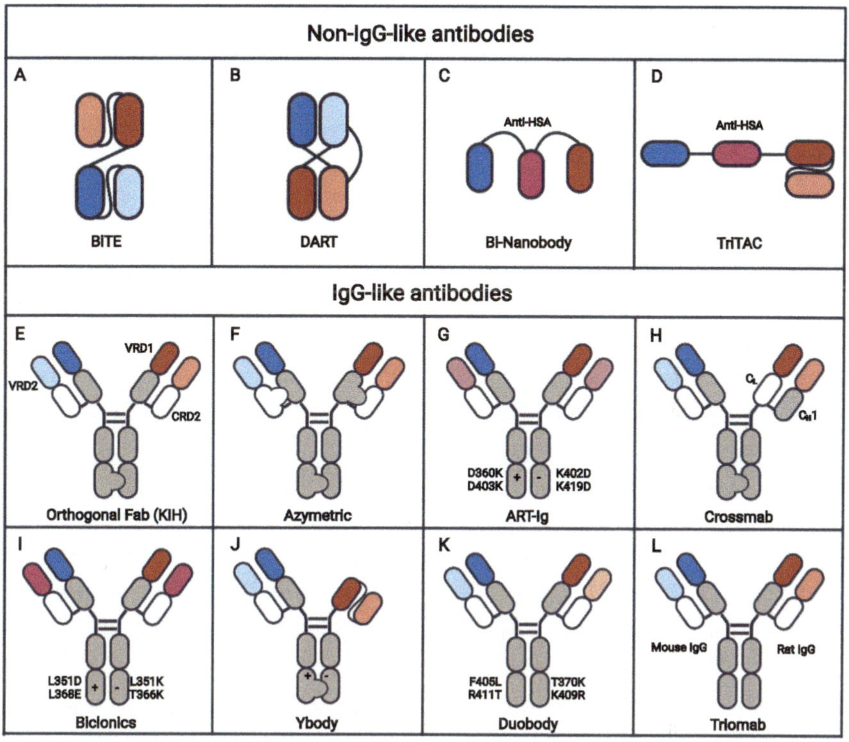

Fig. 2 Currently common platforms for the expression of Bi-specific Antibodies. Bi-specific antibodies can be expressed in the form of a loose connection of antigen-binding domains lacking an Fc-part (non-IgG-like platforms, A–D) or in forms of a more classical antibody structure with FC-part (IgG-like platforms, E–L). The bispecific T cell engager (BiTE) platform (A) consists of a protein encompassing two scFv, one targeting CD3 and another specific for a tumor-associated antigen (TAA), connected via a linker. In contrast, the Dual-Affinity *Re*-Targeting (DART) platform (B) is based on 2 separate polypeptides, consisting of a cognate heavy and light chain variable domain each. These two proteins are assembled and stabilized via a disulfide bridge. The Bi-Nanobody platform (C) connects three nanobody moieties into trivalent molecules by two flexible linkers. Of this trimer, one nanobody is specific for human serum albumin (HSA), in order to extend serum half-life of the construct, while the other two are specific for therapeutic targets, such as tumor antigens. The trispecific T-cell activating construct (TriTAC) (D) is composed of two single-domain antibodies and an scFv. Of the IgG-like platforms, the Orthogonal Fab platform (E) solves heavy-light chain (H-LC) mis-pairing issue by introducing mutations to generate an "orthogonal interface" that enables preferential alignment of the different Fab domains combined with Knob-into-Hole (KiH) heavy-heavy chain (H-HC) heterodimerization method. The Azymetric platform (F) also uses KiH for H-HC heterodimerization and introduces mutations into CH3 to ensure the correct H-LC heterodimerization. In contrast, the Asymmetric Reengineering Technology-immunoglobulin (ART-Ig) platform (G) facilitates H-HC heterodimer formation by introducing electrostatic steering mutations and uses a common, shared light

(Continued)

with the third nanobody targeting human serum albumin (HSA) in order to extend the half-life of the BsAb (Fig. 2C). BI-836880 is a BsAb developed using this platform, which targets vascular endothelial growth factor (VEGF) and angiopoietin-2 (ANG2). It is currently in clinical trial in patients with different solid tumors (NCT02674152). A similar approach is used to generate tri-specific T-cell activating constructs (TriTAC) platform from Harpoon Therapeutics. These are composed of two single-domain antibodies and an scFv which specifically bind to a TAA, HSA, and CD3, respectively (Fig. 2D). Based on this platform, the BsAb HPN424 was developed to target CD3 and PSMA. Early results from clinical trials showed a promising efficacy in patients with metastatic castration-resistant prostate cancer (mCRPC) (Bono et al., 2021).

Compared to non-IgG-like BsAb, IgG-like BsAbs show a substantially longer serum half-life. Furthermore, the IgG-like BsAb tend to display a relatively good solubility and stability. Nevertheless, the IgG-like BsAbs face the problem of the unspecific pairing of the two heavy chains with each other or of the heavy and light chains. A number of different approaches have been developed to address this issue. The predominant approaches applied to ensure a reliable hetero-heavy chain assembly are leucine zipper induced heterodimerization (LUZ-Y) (Wranik et al., 2012), bispecific engagement by antibodies based on T-cell receptor (BEAT) and strand-exchange engineered domain CH3 heterodimers (SEEDbody) approaches. In the leucine zipper induced heterodimerization (LUZ-Y) (Wranik et al., 2012) approach additional leucine zippers are fused to the C terminus of each heavy chain. Following transfection, these leucine zippers then force the newly formed antibodies into a hetero heavy-heavy chain association. Following purification, the leucine zipper can be readily removed. In contrast, in the bispecific engagement by antibodies based on T-cell receptor

Fig. 2—Cont'd chain to ensure correct H-LC pairing. The CrossMab platform (H) overcomes H-LC mispairing by exchanging the CH1 of the heavy chain with the CL of the light chain within the Fab of one half of the BsAb, while correct H-HC heterodimer formation is ensured by typical heterodimer technologies, such as KiH. The Biclonics platform (I) also utilizes a shared, common light chain strategy and electrostatic steering to ensure correct H-HC heterodimer formation. The Y-Body platform (J) is an asymmetric bsAb platform which composes of a Fab-Fc and a scFv-Fc, in this way overcoming H-LC mis-pairing issue. The Duobody platform (K) uses a controlled Fab-arm exchange (cFAE) of two parental antibodies that prefer heterodimer over homodimer formation. Of the Triomab platform (L) heavy and light chain constant regions are of different isotypes, preventing mis-assembly.

(BEAT) (Brinkmann and Kontermann, 2017) approach a hetero heavy chain interface is created, which mimics the natural dimerization of the T-cell surface receptors α and β and thus forces hetero chain formation between two heavy chains. Finally, in the strand-exchange engineered domain heterodimers (SEEDbody) (Davis et al., 2010) approach, complementary human IgG and IgA heavy chain domains are fused with each other on the different heavy chains. Once expressed, these hetero-domains allow for the formation of desired AG/GA heterodimers and prevent the formation of homodimers between AG and GA. Nevertheless, the currently by far most common approach used is the so-called "knobs-into-holes" (KiH) (Huang et al., 2020; Merchant et al., 1998a) approach. In this approach, mutations are introduced in different domains to create either a "knob" structure or a "hole" structure in each heavy chain. These complementary mutations very efficiently prevent homodimerization and promote the heterodimerization of newly expressed antibodies.

Subsequently, heavy-light chain mis-pairing issues had to be overcome. In order to prevent heavy-light chain mis-pairing, several different strategies emerged. For instance, a common light chain approach was used, in which identical light chains are shared by two different heavy chains (Jackman et al., 2010; Krah et al., 2017; Merchant et al., 1998b). Alternatively, similar to heterodimerization methods applied for heavy-heavy chain association, some approaches engineered the contact points between the specific light and heavy chain by structure-based design to enhance correct heavy-light chain association (Bonisch et al., 2017; Igawa et al., 2010; Lewis et al., 2014). Furthermore, some approaches exchanged domains between the heavy and light chain (Schanzer et al., 2014), ensuring that the engineered light chain can only associate to its corresponding domain-swapped heavy chain.

Based on a combination of different of the above named approaches, several companies developed their own platforms to efficiently purify highly pure BsAbs from transfected cells. One example using some of the above-named approaches is the Orthogonal Fab platform Lilly (Lewis et al., 2014). This platform prevents heavy-light chain mis-pairing issues by introducing mutations to generate an "orthogonal interface" that enables preferential alignment of the different Fab domains with correct assembly. Additional KiH mutations in the heavy chain (Fig. 2E) allow for the very efficient and reliable formation of BsAbs. This platform has been used for the construction of several BsAbs. Nevertheless, the BsAb LY3164530, produced by this platform, simultaneously targeting EGFR and c-MET, had completed a phase I trial (NCT012221882) but further development was stopped due to significant toxicities that were observed in several treated patients (Shim, 2020). Nevertheless, more successful appears an antibody constructed based on a similar platform, the so called Azymetric platform by Zymeworks (Von Kreudenstein et al., 2013). This platform also uses a KiH heavy-heavy chain heterodimerization method (Fig. 2F). Based on the Azymetric platform, the BsAb ZW25 was designed to biparatopically bind domain 2 and 4 in the extracellular region of HER2 simultaneously in order to increase avidity and to inhibit HER2/HER3 heterodimer formation. This BsAb demonstrated potent silencing of HER2 signaling and efficient removal of HER2 from the cell surface. In a phase I study (NCT02892123), this BsAb showed promising clinical efficacy (Meric-Bernstam et al., 2021). Similar to the above described KiH approach, also the Asymmetric Reengineering Technology-immunoglobulin (ART-Ig) platform facilitates H-HC heterodimer formation. Nevertheless, in this approach electrostatic steering mutations are introduced in complementary domain interfaces of each heavy chain, providing them with different charges. Consequently, heavy chain homodimers are prevented from forming due to repulsive charge, while heterodimer formation is favored (Fig. 2G). The Emicizumab BsAb by Genentech has been developed based on this platform and was approved in 2018 to treat patients with hemophilia A. Also, the CrossMab from Roche and the Biclonis platform from Merus utilize combinations of the above explained approaches to construct and express BsAbs. The CrossMab platform overcomes heavy-light chain mispairing by exchanging heavy with light chain domains within the Fab of one-half of the BsAb. Combined with another Fc heterodimer technology such as KiH, BsAb can be produced in a rather reliable way (Fig. 2H). There are a considerable number of BsAbs developed by this platform. Current BsAbs in clinical study are for example EM801 (CD3 × BCMA), RG6026 (CD3 × CD20), RG7802 (CD3 × CEA), Vanucizumab (ANG2 × VEGF), and RG7769 (PD-1 × Tim3) (Yoon et al., 2020). In contrast, the Biclonics platform utilizes both a common light chain strategy and electrostatic steering effects which introduces positively charged mutations in one heavy chain and negatively charged mutations in the second heavy chain (Fig. 2I). Based on this platform, MCLA-128, an ADCC enhanced HER2 × HER3 BsAb, is undergoing clinical evaluation for patients with solid tumors harboring Neuroregulin1 (NRG1) fusion (Wu et al., 2021).

Nevertheless, in order to entirely prevent any possibility for mismatching, fundamentally alternative approaches and manufacturing platforms have been developed. Some platforms only use heavy chain heterodimerization methods in which they combine a Fab with an scFv or nanobody to avoid the heavy-light chain mis-pairing issue. For instance, the Y-body platform (Zhou et al., 2012) from YZYBio, combines a Fab-Fc and an scFv-Fc to overcome heavy-light chain (H-LC) mis-pairing. Both Knobs-in-Holes (KiH) and electrostatic steering effects or "salt-bridge" technologies are applied on an Fc-CH3 interface to overcome heavy chain misassembling (Fig. 2J). M802 (anti-HER2 × anti-CD3), developed by this platform, was approved for clinical trials to treat advanced HER2-positive solid tumors (Zhou et al., 2012). Similarly, the Immune-therapy antibody (ITab™) platform (Cui et al., 2016) developed by Shanghai Generon uses two scFv-Fc to overcome heavy-light chain mis-pairing. Based on this platform, A-319, a T-cell activating BsAb designed to target CD19 and CD3 is under clinical evaluation for the treatment of patients with refractory/ relapsed B cell lymphoma (Cui et al., 2016) Finally, a fundamentally different and very successful approach is used by Genmab in the DuoBody technology. The DuoBody technology does not rely on the formation of BsAb in transfected cells but basically mimics the in vivo naturally occurring process of human IgG4 Fab-arm exchange which results in generation of bispecific antibodies (Labrijn et al., 2013; van der Neut Kolfschoten et al., 2007). Two parental IgG1s, each containing destabilizing mutations in their heavy chain domain, are produced separately. Subsequently, the purified antibodies are dissociated and mixed with their pairing partner and then reassembled in vitro. This so-called "controlled Fab-arm exchange" (cFAE) allows for the formation of a wide range of different BsAbs (Fig. 2K). This platform has been licensed-out to many pharmaceutical companies such as BioNTech, Janssen Biotech, Inc., Novo Nordisk and led to several BsAbs in clinical trials such as JNJ-64007957 (CD3 × BCMA), JNJ-63709178 (CD3 × CD123), GEN3013 (CD3 × CD20), JNJ-64407564 (CD3 × GPRC5D) and one, Amivantamab (EGFR × c-Met), recently approved for clinical application.

4. Examples of BsAbs approved for clinical application

Due to the technical success in the assembly of BsAb over the last years, BsAb therapeutics currently represented one of the fastest-growing classes of drugs on the market with four BsAb drugs approved and more than 193 BsAbs in clinical study (Zhang et al., 2021a). Recently, FDA published the Bispecific Antibody Development Programs, the Guidance for Industry

(FDA, 2021) indicative of increasing interest in the production of these reagents. Below, the four BsAbs which have so far been approved by FDA and made it into the market are described. Currently, only three of them are available.

4.1 Catumaxomab (*Removab*, Fresenius Biotech and Trion Pharma)

Approved in 2009 but retracted in 2017, Catumaxomab was a trifunctional rat-mouse hybrid IgG2 monoclonal antibody with two different fragment antigen-binding (Fab) domains. This BsAb bound with one arm to EpCAM on tumor cells and with the other arm to CD3 on T cells *in trans*. Furthermore, via its Fc-part, the BsAb engaged with antigen-presenting cells (Chelius et al., 2010). In this way, the BsAbs formed a trifunctional connection between tumor cells, T-cells and antigen-presenting cells. The generation of this BsAb was based on a Triomab IgG-like platform (Fig. 2L). The Fc segment enhanced an effector response after engaging T-cells and mediated antibody-dependent cell-mediates cytotoxicity (ADCC) against tumoral cells. Catumaxomab was indicated for intraperitoneal treatment of malignant ascites in patients with epithelial cell adhesion molecule (EpCAM)-positive carcinomas (Seimetz, 2011). But, in 2017, it was withdrawn from the market due to the insolvency of the manufacturer of the drug substance. Nevertheless, the company has recently initiated a new phase III study for Catumaxomab to treat patients with advanced gastric carcinoma with peritoneal metastasis (NCT04222114).

4.2 Emicizumab (*HEMLIBRA*, Genentech, Inc.)

Approved in 2018, Emicizumab is a humanized modified asymmetric bispecific IgG4 antibody that binds to blood clotting factor IXa and factor X. Its generation is based on the Art-Ig IgG-like platform (Fig. 2G). It is used to treat patients with hemophilia A, a hereditary bleeding disorder caused by a lack of blood clotting factor VIII. This BsAb mimics the function of the coagulation factor VIII by enhancing the formation of the prothrombinase complex and downstream clot formation (Kitazawa et al., 2017; Uchida et al., 2016). Compared to original factor VIII drug, Emicizumab displays an excellent subcutaneous bioavailability and a relative long half-life in vivo, and therefore currently dominates the market of hemophilia A treatment (Kitazawa et al., 2012).

4.3 Blinatumomab (*Blincyto*, Amgen Inc.)

Approved in 2018, Blinatumomab is a fragment-based BsAb, lacking an Fc region, that targets CD19 and CD3 *in trans* to treat relapsed B-precursor acute lymphoblastic leukemia (ALL) (Goebeler and Bargou, 2016; Löffler et al., 2000). Its generation is based on BiTE non-IgG-like platform (Fig. 2A). The unspecific binding of this BsAb with the TCR receptor on CD8 T-cells induces an antigen-independent cross-linking of the TCR on T -cells, thereby inducing a targeted cellular cytotoxicity toward the tumor cell, leading to the deletion of the tumor cell recognized by the BiTE. Blinatumomab has been the first BsAb drug approved by FDA to treat leukemia. Due to their impressive clinical efficacy (NCT01209286 and NCT01466179), the development of T- and NK-cell bispecific engagers have gained extensive attention and investment by pharmaceutical companies (Tian et al., 2021).

4.4 Amivantamab (*Rybrevant*, Janssen Biotech, Inc.)

Approved in 2021, Amivantamab is a fully-humanized IgG1-based bispecific antibody targeting both epidermal growth factor receptor (EGFR) and mesenchymal-epithelial transition factor (MET). Its generation is based on the Duobody IgG-like platform (Fig. 2K) described above and was shown to bind the EGFR independently of the diverse primary and acquired resistance mutations the EGFR develops in advanced cancers and has been shown to downregulate EGFR expression on tumor cells (Castoldi et al., 2013; Moores et al., 2016). It blocks both EGFR and MET signaling via receptor inactivation. Amivantamab showed substantially enhanced clinical efficacy over EGFR blockade monotreatment and is indicated for the treatment of adult patients with locally advanced or metastatic NSCLC harboring EGFR Exon 20 insertion mutations, whose disease has progressed on, or after platinum-based chemotherapy (Vijayaraghavan et al., 2020). As many tumors overexpress MET in order to escape EGFR blockade (Kalyankrishna and Grandis, 2006), it is assumed that Amivantamab prevents this escape and, consequently, enhances treatment efficacy. Nevertheless, owing to a low-fucose-containing Fc-part, this BsAb is also more adept at inducing ADCC than antibodies with traditional Fc-parts (Grugan et al., 2017; Moores et al., 2016). Hence, a combination of both effects may have contributed to the clinical success of this BsAb (Grugan et al., 2017).

5. Further opportunities for bispecific antibodies on the example of regulatory T-cells

The clinical success of the currently approved BsAbs encouraged the development of even more BsAbs. Furthermore, bsAbs are currently tested as potential therapy in fields like autoimmune and inflammatory disease (Zhao, 2020) or as treatment for infectious diseases, such as SARS-CoV2 (De Gasparo et al., 2021). In this review, as a representative example, we will further focus on the generation of *cis*-binding BsAb, with a special outlook on Tregs targeting in cancer (Table 2).

Regulatory T-cells (Tregs) populations play a pivotal role in preserving an immune balance between self-tolerance and effector responses (Sakaguchi et al., 2010). Their dysfunction causes severe autoimmune disease (Sakaguchi et al., 1995). The concept of suppressor T cells was first described in 1970 by Gershon and Kondo (Gershon and Kondo, 1970). After several years of research and the discovery of CD25 as a Treg marker in 1995, Treg populations became substantially better characterized (Sakaguchi et al., 1995). Currently, Tregs are defined as a highly immunosuppressive subset of CD4+ T cells that dampen excessive immune responses and maintain immune homeostasis by inhibiting effector T cells (Teff) proliferation and cytokine production, hence preventing the development of autoimmune diseases and tissue destruction (Yang et al., 2018). FOXP3 was defined as the hallmark transcription factor of Tregs. Genetic alterations of FOXP3 expression induce Treg cell deficiency, causing severe autoimmune disorders, allergy and inflammation (Bennett et al., 2001; Brunkow et al., 2001; Wildin et al., 2001). Hence, enhancing Treg function is currently seen as an attractive approach for the treatment of autoimmune diseases (Eggenhuizen et al., 2020). On the other hand, Tregs have also been identified as a critical cell type interfering with anti-tumor immune responses and several studies have shown that a high frequency of tumor-infiltrating Tregs in combination with low frequency of T effector cells, resulting in a low Teff/Treg ratio, is correlated with a poor prognosis for cancer patients and tumor immunotherapy (Sato et al., 2005; Shang et al., 2015). In particular, in the context of CTLA-4 and PD-1 immune checkpoint inhibitor (ICI) treatments, it became apparent that Tregs play an important role in ICI resistance (Jie et al., 2015; Maj et al., 2017). Therefore, there is a need to find Treg targeting strategies. However, an exclusive Treg cell surface marker is currently not known.

Table 2 BsAb antibodies generated against Tregs.

Reagent trademark	Compounds (administration)	Indications	NCT (phase)/status	Sponsor	Ref.
ATOR-1015	CTLA-4 × OX40 bispecific (iv)	Solid tumors and neoplasms	NCT03782467 (I)/completed	Alligator Bioscience	Kvarnhammar et al. (2019)
ATOR-1144	CTLA-4 × GITR	Solid tumors and hematological malignancies	–	Alligator Bioscience	Fritzell et al. (2019)
KY1055	ICOS × PD-L1		–	Kymab	Sainson et al. (2018)

Several monoclonal antibodies described to specifically target intra-tumoral Tregs in mice, failed to be effective in a clinical setting. This failure is based in part on the difference in biology of these receptors between species. In mice, anti-CTLA-4 antibodies preferentially deplete tumor-infiltrating Tregs by activating FcγR binding (Selby et al., 2013; Simpson et al., 2013). In contrast, in humans, anti-CTLA-4 treatment shows controversial results: a reduction of Treg intratumorally but collateral effects in secondary lymphoid organs (Arce Vargas et al., 2018; Ha et al., 2019). Other monoclonal antibodies have been described to target Tregs in mice but fail to show effects in clinical trials. For instance, Daclizumab (an anti-CD25 mAb aimed to target Tregs) had to be withdrawn from the market for MS treatment following induction of severe liver failure in some treated patients (Stettner et al., 2019).

Therefore, it has become a rather appealing approach to target specifically intra-tumoral Treg populations with BsAb with the aim to enhance the specificity of Treg targeting with an appropriate combination of two independent antigens expressed on Treg cell surfaces.

5.1 Regulatory T-cell expressed cell surface markers

Tregs express a number of different cell surface markers. Most of these are shared by other activated T-cells. Nevertheless, the most promising approaches for BsAbs targeting specifically Tregs have been focused on a combination of the following receptors:

- **CTLA-4** (CD152): CTLA-4 is an inhibitory receptor relative of the T cell costimulatory molecule CD28 (Lenschow et al., 1996) However, in contrast to CD28, CTLA-4 suppresses T cell responses upon binding of CD80/CD86 (Pentcheva-Hoang et al., 2004). CTLA-4 is an activation marker transiently expressed on all activated T-cells, but it is constitutively expressed on the Treg cell surface (Takahashi et al., 2000). Mice selectively deficient in CTLA-4 in Treg cells develop systemic lymphoproliferation and fatal autoimmune disease, indicative of its importance for Treg-mediated maintenance of immunological self-tolerance (Wing et al., 2008).
- **GITR** (TNFRSF18/CD357): glucocorticoid-induced tumor necrosis factor receptor-related protein belongs to the costimulatory receptors tumor necrosis factor receptor superfamily (TNFRsf). Also GITR is an activation marker transiently expressed to some extent on all activated

T-cells. Nevertheless, GITR is crucial during Treg maturation and mature Tregs highly express it on its surface (Mahmud et al., 2014). It was shown that anti-GITR Ab abrogates the suppressive activity of Tregs (McHugh et al., 2002; Shimizu et al., 2002) and that GITR triggering costimulates Tregs (Ronchetti et al., 2004).

- **OX-40** (TNFRSF4/CD132): OX-40 belongs to the costimulatory receptors TNFRsf. Similar to GITR, OX-40 is transiently expressed by recently activated T-cells. In Tregs, OX-40 mediated costimulation is important for Treg proliferation and, in contrast to effector T-cells, its expression is maintained in Tregs (Mahmud et al., 2014).
- **CD25**: There is a constitutively high expression of the IL-2 receptor CD25 on Tregs, allowing Tregs to respond to IL-2 as a requirement for their growth and survival. IL2 binding facilitates immune tolerance in Tregs (Malek, 2008) and upregulates FOXP3 expression and drive their immunosuppressive phenotype. Due to a lack of Treg function, IL-2 deficient mice die due to the early onset of auto-immune diseases (Zeiser and Negrin, 2008). However, CD25 is also a critical activation marker expressed on activated T-cells. Expression level on activated T-cells is lower than on Tregs, but it mediates essential IL-2 induced signals that facilitate effector T-cell survival and proliferation.
- **PD-1** (CD279): PD-1 is expressed by T and B cells and has a role in regulating immune responses. It is an immune checkpoint which down-regulates the immune system. In this way, it plays an essential role in the regulation between protective immunity and immunopathology. Consequently, PD-1 deficient mice are prone to develop auto-immune diseases. Its expression on Tregs is critical to maintaining the immuno-suppressive activity of these cells (Sharpe and Pauken, 2018).
- **CCR4** (CD194): CCR4 is a chemokine receptor expressed on cutaneous T-cells. However, also recently activated, effector Treg cells express elevated levels of it, making it an attractive target for Treg specific depletion (Kurose et al., 2015).

Taken together, none of the above-mentioned cell surface markers is uniquely expressed on Tregs. Thus, targeting either of these cell surface markers using monoclonal antibodies has been associated with sometimes rather severe side effects. Therefore, several BsAbs candidates have been designed and undergone first clinical trials. In the following, the currently generated BsAbs against Tregs which are being developed and under study are described.

6. Current bi-specific antibodies targeting regulatory T-cells

6.1 ATOR-1015

ATOR-1015 is a human CTLA-4 × OX40 bispecific IgG1 antibody from Alligator Bioscience. Its BsAb format consists of an appended IgG (anti-OX-40) with single-chain variable fragments (scFv) (anti-CTLA-4) fused to a light chain. CTLA-4 and OX40 are both expressed on activated T cells, however, are expressed to a substantially higher level on Tregs in the tumor microenvironment (TME) (Arce Vargas et al., 2018; Montler et al., 2016). Hence, this antibody was designed to target and deplete Tregs in the TME. It combines a simultaneous blockade of CTLA-4 and an agonistic OX40 binding. The use of an IgG1 isotype enhances ADCC, and this BsAb shows superior ADCC activity compared to the combination of monoclonal antibodies (Kvarnhammar et al., 2019). Thanks to its encouraging results in preclinical studies (Yachnin et al., 2020), a first-in-human phase I clinical trial was performed, indicated for solid tumors and neoplasms (NCT03782467).

6.2 ATOR-1144

Similar to ATOR-1015, also ATOR-1144 is a humanized bispecific IgG1 antibody. Its BsAb format consists of an appended IgG (anti-GITR) with single-chain scFv (anti-CTLA-4) fused to the light chain. CTLA-4 and GITR are upregulated on activated T cells, however, yet again, to a substantially higher level on Tregs in the TME (Pedroza-Gonzalez et al., 2013). In order to target and deplete Tregs in TME a simultaneous blockade of CTLA-4 was combined with GITR agonistic binding. ATOR-1144 was developed for treatment of solid tumors and hematological malignancies. In preclinical testing, ATOR-1144 showed Treg depletion via ADCC and the activation of effector T and NK cells within the tumor (Fritzell et al., 2019; Millan-Perez Pena et al., 2020).

6.3 KY1055

KY1055 is a human ICOS × PD-L1 bispecific IgG1 antibody produced by Kymab. KY1055 is based on a human IgG1 format that consists of fully human Fab arms targeting ICOS and a modified Fc (Fcab) targeting PD-L1. ICOS is a co-stimulatory molecule transiently expressed on activated T cells and constitutively by Tregs in the TME (Nagase et al., 2017). Via a simultaneous PD-L1 and ICOS binding, KY1055 showed higher anti-tumor activity in several syngeneic pre-clinical tumor models compared to the treatment combining both monoclonal antibodies. In these models, KY1055 depleted Tregs and improved Teff/Treg ratio (Sainson et al., 2018).

7. Concluding remarks

The above mentioned BsAb appear rather promising in specifically targeting intra-tumoral Treg populations. All Treg targeting BsAbs for tumor malignancies treatment currently share one common goal: to block or deplete Tregs, specifically within the TME. Given the increased specificity of a combination of two selective targets, the use of BsAbs with isotypes that bind and activate stimulating Fc receptors may potentially have fewer side effects than the use of similar monoclonal antibodies. The BsAbs may trigger an effector response specifically within the TME and might hence enhance the efficacy of treatment. For example, high ADCC with anti-CTLA-4 mAb achieves predominant depletion of Treg cells without many off-target effects on Teff cells as the level of CTLA-4 expression on Treg cells is substantially higher on Tregs than on Teff cells (Ha et al., 2019). Nevertheless, developing a BsAb with increased selectivity over two individual mAbs turned out to be rather challenging. In particular, it turned out to be very difficult to lower the binding affinity of an individual antibody moiety sufficiently to prevent unspecific binding to cells that express one antigen only, while at the same time achieving the specific binding of the BsAb to cells that express both antigens. Introducing mutations in the target binding site of an antibody is often complicated and unpredictable, easily leading to entire loss of binding rather than in just lowering it. Furthermore, in particular the selective expression levels of specific receptors expressed on the target cell population is critical for the selection of relevant BsAbs targets (Chen et al., 2011; Reusch et al., 2015). However, the exact prediction of cell surface expression of each target on different cell types within distinct organs and places within the body of patients remains also rather unpredictable. This aspect always raises the possibility of inducing unexpected side-effects in treated patients during clinical trials. Therefore, a number of formidable obstacles remain for BsAb prior to us being able to explore the full potential of their efficacy in clinical settings. Nevertheless, BsAbs are becoming more and more relevant for medical scientists due to the newly provided insights into the huge impact these antibodies can have. This provides new opportunities in the treatment of various diseases. These could include even rather unexpected diseases, as for instance, most recently, BsAbs have shown surprisingly promising efficacy in the treatment of SARS-CoV-2 infected patients (De Gasparo et al., 2021).

Thus, taken together, BsAb may have the great potential to improve treatment efficacy and to reduce side effects in treated patients. This may then open entirely novel treatment opportunities for diseases, such as cancer or autoimmune diseases. Nevertheless, for each BsAb also, daunting challenges may have to be overcome prior to any planned clinical rollout.

Funding

This work was supported by the European Union's Horizon 2020 research and innovation programme under the Marie Skłodowska-Curie grant agreement [grant numbers 765,394, 2018].

Disclosure statement

The authors declare no competing interests. Figures created by BioRender.

References

Arce Vargas, F., Furness, A.J.S., Litchfield, K., Joshi, K., Rosenthal, R., Ghorani, E., Solomon, I., Lesko, M.H., Ruef, N., Roddie, C., Henry, J.Y., Spain, L., Ben Aissa, A., Georgiou, A., Wong, Y.N.S., Smith, M., Strauss, D., Hayes, A., Nicol, D., O'Brien, T., Martensson, L., Ljungars, A., Teige, I., Frendeus, B., Melanoma, T.R., Renal, T.R., Consortia, T.R.L., Pule, M., Marafioti, T., Gore, M., Larkin, J., Turajlic, S., Swanton, C., Peggs, K.S., Quezada, S.A., 2018. Fc effector function contributes to the activity of human anti-CTLA-4 antibodies. Cancer Cell 33, 649–663. e644.

Bennett, C.L., Christie, J., Ramsdell, F., Brunkow, M.E., Ferguson, P.J., Whitesell, L., Kelly, T.E., Saulsbury, F.T., Chance, P.F., Ochs, H.D., 2001. The immune dysregulation, polyendocrinopathy, enteropathy, X-linked syndrome (IPEX) is caused by mutations of FOXP3. Nat. Genet. 27, 20–21.

Bonisch, M., Sellmann, C., Maresch, D., Halbig, C., Becker, S., Toleikis, L., Hock, B., Ruker, F., 2017. Novel CH1:CL interfaces that enhance correct light chain pairing in heterodimeric bispecific antibodies. Protein Eng. Des. Sel. 30, 685–696.

Bono, J.S.D., Fong, L., Beer, T.M., Gao, X., Geynisman, D.M., Burris III, H.A., Strauss, J.F., Courtney, K.D., Quinn, D.I., VanderWeele, D.J., Yaron, Y., Law, C.-L., Stein, M.N., 2021. Results of an ongoing phase 1/2a dose escalation study of HPN424, a tri-specific half-life extended PSMA-targeting T-cell engager, in patients with metastatic castration-resistant prostate cancer (mCRPC). J. Clin. Oncol. 39, 5013.

Brinkmann, U., Kontermann, R.E., 2017. The making of bispecific antibodies. MAbs 9, 182–212.

Brunkow, M.E., Jeffery, E.W., Hjerrild, K.A., Paeper, B., Clark, L.B., Yasayko, S.A., Wilkinson, J.E., Galas, D., Ziegler, S.F., Ramsdell, F., 2001. Disruption of a new forkhead/winged-helix protein, scurfin, results in the fatal lymphoproliferative disorder of the scurfy mouse. Nat. Genet. 27, 68–73.

Byrne, H., Conroy, P.J., Whisstock, J.C., O'Kennedy, R.J., 2013. A tale of two specificities: bispecific antibodies for therapeutic and diagnostic applications. Trends Biotechnol. 31, 621–632.

Castoldi, R., Ecker, V., Wiehle, L., Majety, M., Busl-Schuller, R., Asmussen, M., Nopora, A., Jucknischke, U., Osl, F., Kobold, S., Scheuer, W., Venturi, M., Klein, C., Niederfellner, G., Sustmann, C., 2013. A novel bispecific EGFR/met antibody blocks tumor-promoting phenotypic effects induced by resistance to EGFR inhibition and has potent antitumor activity. Oncogene 32, 5593–5601.

Chelius, D., Ruf, P., Gruber, P., Ploscher, M., Liedtke, R., Gansberger, E., Hess, J., Wasiliu, M., Lindhofer, H., 2010. Structural and functional characterization of the trifunctional antibody catumaxomab. MAbs 2, 309–319.

Chen, K.-J., Lin, S.-Z., Zhou, L., Xie, H.-Y., Zhou, W.-H., Taki-Eldin, A., Zheng, S.-S., 2011. Selective recruitment of regulatory T cell through CCR6-CCL20 in hepatocellular carcinoma fosters tumor progression and predicts poor prognosis. PLoS One 6, e24671.

Chen, A.Y., Wolchok, J.D., Bass, A.R., 2021. TNF in the era of immune checkpoint inhibitors: friend or foe? Nat. Rev. Rheumatol. 17, 213–223.

Cui, Y., Huang, Z., Chen, H., Zhang, X., Qi, B., Yan, X., 2016. Multispecific Fab Fusion Proteins and Use Thereof. Itabmed Hk Ltd, United States.

Davis, J.H., Aperlo, C., Li, Y., Kurosawa, E., Lan, Y., Lo, K.M., Huston, J.S., 2010. SEEDbodies: fusion proteins based on strand-exchange engineered domain (SEED) CH3 heterodimers in an fc analogue platform for asymmetric binders or immunofusions and bispecific antibodies. Protein Eng. Des. Sel. 23, 195–202.

De Gasparo, R., Pedotti, M., Simonelli, L., Nickl, P., Muecksch, F., Cassaniti, I., Percivalle, E., Lorenzi, J.C.C., Mazzola, F., Magri, D., Michalcikova, T., Haviernik, J., Honig, V., Mrazkova, B., Polakova, N., Fortova, A., Tureckova, J., Iatsiuk, V., Di Girolamo, S., Palus, M., Zudova, D., Bednar, P., Bukova, I., Bianchini, F., Mehn, D., Nencka, R., Strakova, P., Pavlis, O., Rozman, J., Gioria, S., Sammartino, J.C., Giardina, F., Gaiarsa, S., Pan-Hammarstrom, Q., Barnes, C.O., Bjorkman, P.J., Calzolai, L., Piralla, A., Baldanti, F., Nussenzweig, M.C., Bieniasz, P.D., Hatziioannou, T., Prochazka, J., Sedlacek, R., Robbiani, D.F., Ruzek, D., Varani, L., 2021. Bispecific IgG neutralizes SARS-CoV-2 variants and prevents escape in mice. Nature 593, 424–428.

DiLillo, D.J., Olson, K., Mohrs, K., Meagher, T.C., Bray, K., Sineshchekova, O., Startz, T., Kuhnert, J., Retter, M.W., Godin, S., Sharma, P., Delfino, F., Lin, J., Smith, E., Thurston, G., Kirshner, J.R., 2021. A BCMAxCD3 bispecific T cell-engaging antibody demonstrates robust antitumor efficacy similar to that of anti-BCMA CAR T cells. Blood Adv. 5, 1291–1304.

Eggenhuizen, P.J., Ng, B.H., Ooi, J.D., 2020. Treg enhancing therapies to treat autoimmune diseases. Int. J. Mol. Sci. 21.

Fan, G., Wang, Z., Hao, M., Li, J., 2015. Bispecific antibodies and their applications. J. Hematol. Oncol. 8, 130.

FDA, 2021. Bispecific Antibody Development Programs Guidance for Industry. Services, U. S. D. o. H. a. H., Administration, F. a. D., (CDER), C. f. D. E. a. R., and (CBER), C. f. B. E. a. R, Ed.

Fritzell, S., Levin, M., Åberg, I., Johansson, M., Winnerstam, M., Smith, K.E., Ellmark, P., Furebring, C., Norlén, P., Kvarnhammar, A., 2019. Abstract 4077: ATOR-1144 is a tumor-directed CTLA-4 x GITR bispecific antibody that acts by depleting Tregs and activating effector T cells and NK cells. Cancer Res. 79, 4077.

Gershon, R.K., Kondo, K., 1970. Cell interactions in the induction of tolerance: the role of thymic lymphocytes. Immunology 18, 723–737.

Goebeler, M.E., Bargou, R., 2016. Blinatumomab: a CD19/CD3 bispecific T cell engager (BiTE) with unique anti-tumor efficacy. Leuk. Lymphoma 57, 1021–1032.

Grugan, K.D., Dorn, K., Jarantow, S.W., Bushey, B.S., Pardinas, J.R., Laquerre, S., Moores, S.L., Chiu, M.L., 2017. Fc-mediated activity of EGFR x c-met bispecific antibody JNJ-61186372 enhanced killing of lung cancer cells. MAbs 9, 114–126.

Ha, D., Tanaka, A., Kibayashi, T., Tanemura, A., Sugiyama, D., Wing, J.B., Lim, E.L., Teng, K.W.W., Adeegbe, D., Newell, E.W., Katayama, I., Nishikawa, H., Sakaguchi, S., 2019. Differential control of human Treg and effector T cells in tumor immunity by fc-engineered anti-CTLA-4 antibody. Proc. Natl. Acad. Sci. U. S. A. 116, 609–618.

Hosseini, S.S., Khalili, S., Baradaran, B., Bidar, N., Shahbazi, M.A., Mosafer, J., Hashemzaei, M., Mokhtarzadeh, A., Hamblin, M.R., 2021. Bispecific monoclonal antibodies for targeted immunotherapy of solid tumors: recent advances and clinical trials. Int. J. Biol. Macromol. 167, 1030–1047.

Huang, S., Segues, A., Hulsik, D.L., Zaiss, D.M., Sijts, A., van Duijnhoven, S.M.J., van Elsas, A., 2020. A novel efficient bispecific antibody format, combining a conventional antigen-binding fragment with a single domain antibody, avoids potential heavy-light chain mis-pairing. J. Immunol. Methods 483, 112811.

Igawa, T., Tsunoda, H., Kikuchi, Y., Yoshida, M., Tanaka, M., Koga, A., Sekimori, Y., Orita, T., Aso, Y., Hattori, K., Tsuchiya, M., 2010. VH/VL interface engineering to promote selective expression and inhibit conformational isomerization of thrombopoietin receptor agonist single-chain diabody. Protein Eng. Des. Sel. 23, 667–677.

Jackman, J., Chen, Y., Huang, A., Moffat, B., Scheer, J.M., Leong, S.R., Lee, W.P., Zhang, J., Sharma, N., Lu, Y., Iyer, S., Shields, R.L., Chiang, N., Bauer, M.C., Wadley, D., Roose-Girma, M., Vandlen, R., Yansura, D.G., Wu, Y., Wu, L.C., 2010. Development of a two-part strategy to identify a therapeutic human bispecific antibody that inhibits IgE receptor signaling. J. Biol. Chem. 285, 20850–20859.

Jie, H.B., Schuler, P.J., Lee, S.C., Srivastava, R.M., Argiris, A., Ferrone, S., Whiteside, T.L., Ferris, R.L., 2015. CTLA-4(+) regulatory T cells increased in Cetuximab-treated head and neck Cancer patients suppress NK cell cytotoxicity and correlate with poor prognosis. Cancer Res. 75, 2200–2210.

Kalyankrishna, S., Grandis, J.R., 2006. Epidermal growth factor receptor biology in head and neck cancer. J. Clin. Oncol. 24, 2666–2672.

Kitazawa, T., Igawa, T., Sampei, Z., Muto, A., Kojima, T., Soeda, T., Yoshihashi, K., Okuyama-Nishida, Y., Saito, H., Tsunoda, H., Suzuki, T., Adachi, H., Miyazaki, T., Ishii, S., Kamata-Sakurai, M., Iida, T., Harada, A., Esaki, K., Funaki, M., Moriyama, C., Tanaka, E., Kikuchi, Y., Wakabayashi, T., Wada, M., Goto, M., Toyoda, T., Ueyama, A., Suzuki, S., Haraya, K., Tachibana, T., Kawabe, Y., Shima, M., Yoshioka, A., Hattori, K., 2012. A bispecific antibody to factors IXa and X restores factor VIII hemostatic activity in a hemophilia a model. Nat. Med. 18, 1570–1574.

Kitazawa, T., Esaki, K., Tachibana, T., Ishii, S., Soeda, T., Muto, A., Kawabe, Y., Igawa, T., Tsunoda, H., Nogami, K., Shima, M., Hattori, K., 2017. Factor VIIIa-mimetic cofactor activity of a bispecific antibody to factors IX/IXa and X/Xa, emicizumab, depends on its ability to bridge the antigens. Thromb. Haemost. 117, 1348–1357.

Klein, C., Sustmann, C., Thomas, M., Stubenrauch, K., Croasdale, R., Schanzer, J., Brinkmann, U., Kettenberger, H., Regula, J.T., Schaefer, W., 2012. Progress in overcoming the chain association issue in bispecific heterodimeric IgG antibodies. MAbs 4, 653–663.

Krah, S., Schroter, C., Eller, C., Rhiel, L., Rasche, N., Beck, J., Sellmann, C., Gunther, R., Toleikis, L., Hock, B., Kolmar, H., Becker, S., 2017. Generation of human bispecific common light chain antibodies by combining animal immunization and yeast display. Protein Eng. Des. Sel. 30, 291–301.

Kurose, K., Ohue, Y., Wada, H., Iida, S., Ishida, T., Kojima, T., Doi, T., Suzuki, S., Isobe, M., Funakoshi, T., Kakimi, K., Nishikawa, H., Udono, H., Oka, M.,

Ueda, R., Nakayama, E., 2015. Phase Ia study of FoxP3 + CD4 Treg depletion by infusion of a humanized anti-CCR4 antibody, KW-0761, in cancer patients. Clin. Cancer Res. 21, 4327–4336.

Kvarnhammar, A.M., Veitonmaki, N., Hagerbrand, K., Dahlman, A., Smith, K.E., Fritzell, S., von Schantz, L., Thagesson, M., Werchau, D., Smedenfors, K., Johansson, M., Rosen, A., Aberg, I., Winnerstam, M., Nyblom, E., Barchan, K., Furebring, C., Norlen, P., Ellmark, P., 2019. The CTLA-4 x OX40 bispecific antibody ATOR-1015 induces anti-tumor effects through tumor-directed immune activation. J. Immunother. Cancer 7, 103.

Labrijn, A.F., Meesters, J.I., de Goeij, B.E., van den Bremer, E.T., Neijssen, J., van Kampen, M.D., Strumane, K., Verploegen, S., Kundu, A., Gramer, M.J., van Berkel, P.H., van de Winkel, J.G., Schuurman, J., Parren, P.W., 2013. Efficient generation of stable bispecific IgG1 by controlled fab-arm exchange. Proc. Natl. Acad. Sci. U. S. A. 110, 5145–5150.

Labrijn, A.F., Janmaat, M.L., Reichert, J.M., Parren, P., 2019. Bispecific antibodies: a mechanistic review of the pipeline. Nat. Rev. Drug Discov. 18, 585–608.

Lenschow, D.J., Walunas, T.L., Bluestone, J.A., 1996. CD28/B7 system of T cell costimulation. Annu. Rev. Immunol. 14, 233–258.

Lewis, S.M., Wu, X., Pustilnik, A., Sereno, A., Huang, F., Rick, H.L., Guntas, G., Leaver-Fay, A., Smith, E.M., Ho, C., Hansen-Estruch, C., Chamberlain, A.K., Truhlar, S.M., Conner, E.M., Atwell, S., Kuhlman, B., Demarest, S.J., 2014. Generation of bispecific IgG antibodies by structure-based design of an orthogonal fab interface. Nat. Biotechnol. 32, 191–198.

Loffler, A., Kufer, P., Lutterbuse, R., Zettl, F., Daniel, P.T., Schwenkenbecher, J.M., Riethmuller, G., Dorken, B., Bargou, R.C., 2000. A recombinant bispecific single-chain antibody, CD19xCD3, induces rapid and high lymphoma-directed cytotoxicity by unstimulated T lymphocytes. Blood 95, 2098–2103.

Löffler, A., Kufer, P., Lutterbüse, R., Zettl, F., Daniel, P.T., Schwenkenbecher, J.M., Riethmüller, G., Dörken, B., Bargou, R.C., 2000. A recombinant bispecific single-chain antibody, CD19 × CD3, induces rapid and high lymphoma-directed cytotoxicity by unstimulated T lymphocytes. Blood 95, 2098–2103.

Mahmud, S.A., Manlove, L.S., Schmitz, H.M., Xing, Y., Wang, Y., Owen, D.L., Schenkel, J.M., Boomer, J.S., Green, J.M., Yagita, H., Chi, H., Hogquist, K.A., Farrar, M.A., 2014. Costimulation via the tumor-necrosis factor receptor superfamily couples TCR signal strength to the thymic differentiation of regulatory T cells. Nat. Immunol. 15, 473–481.

Maj, T., Wang, W., Crespo, J., Zhang, H., Wang, W., Wei, S., Zhao, L., Vatan, L., Shao, I., Szeliga, W., Lyssiotis, C., Liu, J.R., Kryczek, I., Zou, W., 2017. Oxidative stress controls regulatory T cell apoptosis and suppressor activity and PD-L1-blockade resistance in tumor. Nat. Immunol. 18, 1332–1341.

Malek, T.R., 2008. The biology of interleukin-2. Annu. Rev. Immunol. 26, 453–479.

Mazor, Y., Hansen, A., Yang, C., Chowdhury, P.S., Wang, J., Stephens, G., Wu, H., Dall'Acqua, W.F., 2015. Insights into the molecular basis of a bispecific antibody's target selectivity. MAbs 7, 461–469.

McHugh, R.S., Whitters, M.J., Piccirillo, C.A., Young, D.A., Shevach, E.M., Collins, M., Byrne, M.C., 2002. CD4(+)CD25(+) immunoregulatory T cells: gene expression analysis reveals a functional role for the glucocorticoid-induced TNF receptor. Immunity 16, 311–323.

Merchant, A.M., Zhu, Z., Yuan, J.Q., Goddard, A., Adams, C.W., Presta, L.G., Carter, P., 1998a. An efficient route to human bispecific IgG. Nat. Biotechnol. 16, 677–681.

Merchant, A.M., Zhu, Z., Yuan, J.Q., Goddard, A., Adams, C.W., Presta, L.G., Carter, P., 1998b. An efficient route to human bispecific IgG. Nat. Biotechnol. 16, 677–681.

Meric-Bernstam, F., Hamilton, E.P., Beeram, M., Hanna, D.L., El-Khoueiry, A.B., Kang, Y.-K., Lee, K.W., Lee, J., Rha, S.Y., Chaves, J.M., Oh, D.-Y., Goodwin, R.A., Ajani, J.A., Gray, T., Woolery, J., Elimova, E., 2021. Zanidatamab (ZW25) in HER2-expressing gastroesophageal adenocarcinoma (GEA): results from a phase I study. J. Clin. Oncol. 39, 164.

Millan-Perez Pena, L., Martin, P.S., Herrera-Camacho, I., Bandala, C., Anaya-Ruiz, M., 2020. Colon carcinoma treatment using bispecific anti-GITR/CTLA-4 antibodies: a patent evaluation of WO2018091739. Expert Opin. Ther. Pat. 30, 307–311.

Milstein, C., Cuello, A.C., 1983. Hybrid hybridomas and their use in immunohistochemistry. Nature 305, 537–540.

Montler, R., Bell, R.B., Thalhofer, C., Leidner, R., Feng, Z., Fox, B.A., Cheng, A.C., Bui, T.G., Tucker, C., Hoen, H., Weinberg, A., 2016. OX40, PD-1 and CTLA-4 are selectively expressed on tumor-infiltrating T cells in head and neck cancer. Clin. Transl. Immunol. 5, e70.

Moores, S.L., Chiu, M.L., Bushey, B.S., Chevalier, K., Luistro, L., Dorn, K., Brezski, R.J., Haytko, P., Kelly, T., Wu, S.J., Martin, P.L., Neijssen, J., Parren, P.W., Schuurman, J., Attar, R.M., Laquerre, S., Lorenzi, M.V., Anderson, G.M., 2016. A novel bispecific antibody targeting EGFR and cMet is effective against EGFR inhibitor-resistant lung tumors. Cancer Res. 76, 3942–3953.

Nagase, H., Takeoka, T., Urakawa, S., Morimoto-Okazawa, A., Kawashima, A., Iwahori, K., Takiguchi, S., Nishikawa, H., Sato, E., Sakaguchi, S., Mori, M., Doki, Y., Wada, H., 2017. ICOS(+) Foxp3(+) TILs in gastric cancer are prognostic markers and effector regulatory T cells associated with Helicobacter pylori. Int. J. Cancer 140, 686–695.

Nisonoff, A., Wissler, F.C., Lipman, L.N., 1960. Properties of the major component of a peptic digest of rabbit antibody. Science 132, 1770–1771.

Pedroza-Gonzalez, A., Verhoef, C., Ijzermans, J.N., Peppelenbosch, M.P., Kwekkeboom, J., Verheij, J., Janssen, H.L., Sprengers, D., 2013. Activated tumor-infiltrating CD4+ regulatory T cells restrain antitumor immunity in patients with primary or metastatic liver cancer. Hepatology 57, 183–194.

Pentcheva-Hoang, T., Egen, J.G., Wojnoonski, K., Allison, J.P., 2004. B7-1 and B7-2 selectively recruit CTLA-4 and CD28 to the immunological synapse. Immunity 21, 401–413.

Reusch, U., Duell, J., Ellwanger, K., Herbrecht, C., Knackmuss, S.H., Fucek, I., Eser, M., McAleese, F., Molkenthin, V., Gall, F.L., Topp, M., Little, M., Zhukovsky, E.A., 2015. A tetravalent bispecific TandAb (CD19/CD3), AFM11, efficiently recruits T cells for the potent lysis of CD19(+) tumor cells. MAbs 7, 584–604.

Rogala, B., Freyer, C.W., Ontiveros, E.P., Griffiths, E.A., Wang, E.S., Wetzler, M., 2015. Blinatumomab: enlisting serial killer T-cells in the war against hematologic malignancies. Expert. Opin. Biol. Ther. 15, 895–908.

Ronchetti, S., Zollo, O., Bruscoli, S., Agostini, M., Bianchini, R., Nocentini, G., Ayroldi, E., Riccardi, C., 2004. GITR, a member of the TNF receptor superfamily, is costimulatory to mouse T lymphocyte subpopulations. Eur. J. Immunol. 34, 613–622.

Sainson, R.C., Parveen, N., Borhis, G., Kosmac, M., Okell, T., Taggart, E., Carvalho, J., McCourt, M., Ali, H., Craig, H., Labokha, A., 2018. Abstract LB-153: KY1055, a novel ICOS-PD-L1 bispecific antibody, efficiently enhances T cell activation and delivers a potent anti-tumour response in vivo. Cancer Res. 78, LB-153.

Sakaguchi, S., Sakaguchi, N., Asano, M., Itoh, M., Toda, M., 1995. Immunologic self-tolerance maintained by activated T cells expressing IL-2 receptor alpha-chains (CD25). Breakdown of a single mechanism of self-tolerance causes various autoimmune diseases. J. Immunol. 155, 1151–1164.

Sakaguchi, S., Miyara, M., Costantino, C.M., Hafler, D.A., 2010. FOXP3+ regulatory T cells in the human immune system. Nat. Rev. Immunol. 10, 490–500.

Sato, E., Olson, S.H., Ahn, J., Bundy, B., Nishikawa, H., Qian, F., Jungbluth, A.A., Frosina, D., Gnjatic, S., Ambrosone, C., Kepner, J., Odunsi, T., Ritter, G., Lele, S., Chen, Y.T., Ohtani, H., Old, L.J., Odunsi, K., 2005. Intraepithelial CD8+ tumor-infiltrating lymphocytes and a high CD8+/regulatory T cell ratio are associated with favorable prognosis in ovarian cancer. Proc. Natl. Acad. Sci. U. S. A. 102, 18538–18543.

Schanzer, J.M., Wartha, K., Croasdale, R., Moser, S., Kunkele, K.P., Ries, C., Scheuer, W., Duerr, H., Pompiati, S., Pollman, J., Stracke, J., Lau, W., Ries, S., Brinkmann, U., Klein, C., Umana, P., 2014. A novel glycoengineered bispecific antibody format for targeted inhibition of epidermal growth factor receptor (EGFR) and insulin-like growth factor receptor type I (IGF-1R) demonstrating unique molecular properties. J. Biol. Chem. 289, 18693–18706.

Seimetz, D., 2011. Novel monoclonal antibodies for cancer treatment: the trifunctional antibody catumaxomab (removab). J. Cancer 2, 309–316.

Selby, M.J., Engelhardt, J.J., Quigley, M., Henning, K.A., Chen, T., Srinivasan, M., Korman, A.J., 2013. Anti-CTLA-4 antibodies of IgG2a isotype enhance antitumor activity through reduction of intratumoral regulatory T cells. Cancer Immunol. Res. 1, 32.

Shang, B., Liu, Y., Jiang, S.J., Liu, Y., 2015. Prognostic value of tumor-infiltrating FoxP3+ regulatory T cells in cancers: a systematic review and meta-analysis. Sci. Rep. 5, 15179.

Sharpe, A.H., Pauken, K.E., 2018. The diverse functions of the PD1 inhibitory pathway. Nat. Rev. Immunol. 18, 153–167.

Shim, H., 2020. Bispecific antibodies and antibody-drug conjugates for cancer therapy: technological considerations. Biomolecules 10, 360.

Shimizu, J., Yamazaki, S., Takahashi, T., Ishida, Y., Sakaguchi, S., 2002. Stimulation of CD25(+)CD4(+) regulatory T cells through GITR breaks immunological self-tolerance. Nat. Immunol. 3, 135–142.

Simpson, T.R., Li, F., Montalvo-Ortiz, W., Sepulveda, M.A., Bergerhoff, K., Arce, F., Roddie, C., Henry, J.Y., Yagita, H., Wolchok, J.D., Peggs, K.S., Ravetch, J.V., Allison, J.P., Quezada, S.A., 2013. Fc-dependent depletion of tumor-infiltrating regulatory T cells co-defines the efficacy of anti-CTLA-4 therapy against melanoma. J. Exp. Med. 210, 1695–1710.

Stettner, M., Gross, C.C., Mausberg, A.K., Pul, R., Junker, A., Baba, H.A., Schulte-Mecklenbeck, A., Wiendl, H., Kleinschnitz, C., Meuth, S.G., 2019. A fatal case of daclizumab-induced liver failure in a patient with MS. Neurol. Neuroimmunol. Neuroinflamm. 6, e539.

Takahashi, T., Tagami, T., Yamazaki, S., Uede, T., Shimizu, J., Sakaguchi, N., Mak, T.W., Sakaguchi, S., 2000. Immunologic self-tolerance maintained by CD25(+)CD4(+) regulatory T cells constitutively expressing cytotoxic T lymphocyte-associated antigen 4. J. Exp. Med. 192, 303–310.

Tian, Z., Liu, M., Zhang, Y., Wang, X., 2021. Bispecific T cell engagers: an emerging therapy for management of hematologic malignancies. J. Hematol. Oncol. 14, 75.

Uchida, N., Sambe, T., Yoneyama, K., Fukazawa, N., Kawanishi, T., Kobayashi, S., Shima, M., 2016. A first-in-human phase 1 study of ACE910, a novel factor VIII-mimetic bispecific antibody, in healthy subjects. Blood 127, 1633–1641.

van der Neut Kolfschoten, M., Schuurman, J., Losen, M., Bleeker, W.K., Martinez-Martinez, P., Vermeulen, E., den Bleker, T.H., Wiegman, L., Vink, T., Aarden, L.A., De Baets, M.H., van de Winkel, J.G., Aalberse, R.C., Parren, P.W., 2007. Anti-inflammatory activity of human IgG4 antibodies by dynamic Fab arm exchange. Science 317, 1554–1557.

Vijayaraghavan, S., Lipfert, L., Chevalier, K., Bushey, B.S., Henley, B., Lenhart, R., Sendecki, J., Beqiri, M., Millar, H.J., Packman, K., Lorenzi, M.V., Laquerre, S., Moores, S.L., 2020. Amivantamab (JNJ-61186372), an Fc enhanced EGFR/cMet bispecific antibody, induces receptor downmodulation and antitumor activity by monocyte/macrophage trogocytosis. Mol. Cancer Ther. 19, 2044.

Von Kreudenstein, T.S., Escobar-Carbrera, E., Lario, P.I., D'Angelo, I., Brault, K., Kelly, J., Durocher, Y., Baardsnes, J., Woods, R.J., Xie, M.H., Girod, P.A., Suits, M.D., Boulanger, M.J., Poon, D.K., Ng, G.Y., Dixit, S.B., 2013. Improving biophysical properties of a bispecific antibody scaffold to aid developability: quality by molecular design. MAbs 5, 646–654.

Vukovic, N., van Elsas, A., Verbeek, J.S., Zaiss, D.M.W., 2021. Isotype selection for antibody-based cancer therapy. Clin. Exp. Immunol. 203, 351–365.

Wildin, R.S., Ramsdell, F., Peake, J., Faravelli, F., Casanova, J.-L., Buist, N., Levy-Lahad, E., Mazzella, M., Goulet, O., Perroni, L., Dagna Bricarelli, F., Byrne, G., McEuen, M., Proll, S., Appleby, M., Brunkow, M.E., 2001. X-linked neonatal diabetes mellitus, enteropathy and endocrinopathy syndrome is the human equivalent of mouse scurfy. Nat. Genet. 27, 18–20.

Wing, K., Onishi, Y., Prieto-Martin, P., Yamaguchi, T., Miyara, M., Fehervari, Z., Nomura, T., Sakaguchi, S., 2008. CTLA-4 control over Foxp3+ regulatory T cell function. Science 322, 271–275.

Wranik, B.J., Christensen, E.L., Schaefer, G., Jackman, J.K., Vendel, A.C., Eaton, D., 2012. LUZ-Y, a novel platform for the mammalian cell production of full-length IgG-bispecific antibodies. J. Biol. Chem. 287, 43331–43339.

Wu, Y., Yi, M., Zhu, S., Wang, H., Wu, K., 2021. Recent advances and challenges of bispecific antibodies in solid tumors. Exp. Hematol. Oncol. 10, 56.

Yachnin, J., Ullenhag, G.J., Carneiro, A., Nielsen, D., Rohrberg, K.S., Kvarnhammar, A.M., Schultz, L., Bågeman, E., Wennersten, C., Russell, C.A., 2020. A first-in-human phase I study in patients with advanced and/or refractory solid malignancies to evaluate the safety of ATOR-1015, a CTLA-4 x OX40 bispecific antibody. J. Clin. Oncol. 38, 3061.

Yang, S., Wang, J., Brand, D.D., Zheng, S.G., 2018. Role of TNF-TNF receptor 2 signal in regulatory T cells and its therapeutic implications. Front. Immunol. 9, 784.

Yoon, A., Lee, S., Lee, S., Lim, S., Park, Y.Y., Song, E., Kim, D.S., Kim, K., Lim, Y., 2020. A novel T cell-engaging bispecific antibody for treating mesothelin-positive solid tumors. Biomol. Ther. 10.

Zeiser, R., Negrin, R.S., 2008. Interleukin-2 receptor downstream events in regulatory T cells: implications for the choice of immunosuppressive drug therapy. Cell Cycle 7, 458–462.

Zhang, Z., Luo, F., Cao, J., Lu, F., Zhang, Y., Ma, Y., Zeng, K., Zhang, L., Zhao, H., 2021a. Anticancer bispecific antibody R&D advances: a study focusing on research trend worldwide and in China. J. Hematol. Oncol. 14, 124.

Zhang, H., Shi, N., Diao, Z., Chen, Y., Zhang, Y., 2021b. Therapeutic potential of TNFalpha inhibitors in chronic inflammatory disorders: past and future. Genes Dis. 8, 38–47.

Zhao, Q., 2020. Bispecific antibodies for autoimmune and inflammatory diseases: clinical progress to date. BioDrugs 34, 111–119.

Zhou, P., Zhang, J., Yan, Y., 2012. Bispecific Antibody. U.S. Patent Ed. WUHAN YZY BIOPHARMA Co. Ltd.

CHAPTER THREE

Applications of trimerbodies in cancer immunotherapy

Marta Compte[a], Laura Sanz[b], and Luis Álvarez-Vallina[c,d,*]
[a]Department of Antibody Engineering, Leadartis S.L., Madrid, Spain
[b]Molecular Immunology Unit, Hospital Universitario Puerta de Hierro Majadahonda, Majadahonda, Madrid, Spain
[c]Cancer Immunotherapy Unit (UNICA), Department of Immunology, Hospital Universitario12 de Octubre, Madrid, Spain
[d]Immuno-Oncology and Immunotherapy Group, Instituto de Investigación Sanitaria 12 de Octubre (imas12), Madrid, Spain
*Corresponding author: e-mail address: lav.imas12@h12o.es

Contents

Abstract

Trimerbodies, with their unique structural and functional properties, are the basis of a new generation of therapeutic antibodies, which due to their small size and plasticity are ideal for the generation of novel biological protein drugs with multiple competitive advantages over conventional full-length monoclonal antibodies. Since their emergence, trimerbodies have been used in preclinical cancer diagnosis and therapy. Trimerbodies are highly adaptable molecules, as they allow target-specific modulation of T cell–mediated anti-tumor immunity to enhance preexisting responses or to generate de novo immune responses. In fact, a tumor-specific humanized 4-1BB-agonistic trimerbody has shown a rather impressive safety and efficacy profile in preclinical studies making it a realistic option for clinical development. Moreover, thanks to the avidity effect they are endowed with considerable therapeutic potential as carriers to deliver cytotoxic payloads to tumors. In addition, molecular imaging studies could benefit from some intermediate-sized trivalent trimerbodies as promising candidates for targeted therapy and tumor imaging.

International Review of Cell and Molecular Biology, Volume 369
ISSN 1937-6448
https://doi.org/10.1016/bs.ircmb.2022.03.008

1. Introduction

Monoclonal antibodies (mAbs) have become a powerful tool for the generation of highly specific molecules with multiple applications in cancer diagnosis and therapy (Sanz et al., 2005). As of November, 2021, more than one hundred mAbs have been approved by regulatory agencies for therapeutic use, half of them in oncological patients (2021). Therapeutic full-length IgG mAbs have direct (target-specific) and/or indirect (antibody-specific) modes of action. Direct effects depend on the antigen and epitope recognized, and the antigen–antibody interaction can modulate biological responses, interfering with tumor growth signals and possibly triggering apoptosis (Scott et al., 2012). Indirect effects include induction of cytotoxic mechanisms, which are triggered by interactions of the fragment crystallizable (Fc) domain with Fc receptors (FcRs) and/or complement factors (Scott et al., 2012). Unfortunately, the activation of Fc-dependent functions by therapeutic mAbs is facing various limitations, such as the impact of FcR polymorphisms on clinical responses (Reichert and Dhimolea, 2012), and the interaction with inhibitory FcRs (Nimmerjahn and Ravetch, 2007).

Furthermore, it has been found that solid neoplasms are particularly resistant to full-length mAb-based therapies. In fact, only a minor fraction of systemically delivered IgGs accumulates in the tumor (Beckman et al., 2007; Cuesta et al., 2010). Antibody uptake by tumors relies on a complex equilibrium between favorable pharmacokinetics, and effective tumor penetration, and retention. Several parameters, such as molecular weight, specificity, affinity, valency, antigen density, and half-life extension should be carefully considered in order to design the optimal antibody-based targeting agent (Xenaki et al., 2017). In response to these shortcomings, a wide range of engineered antibody formats have been generated (Cuesta et al., 2010).

2. Next-generation antibodies

Early efforts to enhance tumor penetration were focused on smaller antibody derivatives, such as the single-chain variable fragment (scFv) consisting of an interconnected variable heavy chain (VH) and variable light chain (VL). The scFv antibodies (25–30 kDa) display improved pharmacokinetics over IgG molecules for tumor penetration (Cuesta et al., 2010). However, scFvs have rapid blood clearance, as well as fast off-rates and poor tumor retention time, due to their monovalent binding (Cuesta et al., 2010). Several strategies have been proposed to further prolong the serum half-life of antibody fragments (Kontermann, 2009; Mandrup and Howard, 2021). The chemical coupling of polyethylene glycol (PEG) is a well-established method to improve the serum half-life of proteins by increasing the molecular size and thus reducing glomerular filtration, and several pEGylated therapeutic proteins have been approved by regulatory agencies, including an anti-tumor necrosis factor (TNF) Fab' fragment (Certolizumab pegol) (Kontermann, 2009). However, sometimes PEGylation leads to partial inactivation or reduced affinity of the antibody, and different approaches are being explored, such as fusion with human serum albumin (Kontermann, 2016) or variants mutated for increased affinity for the neonatal Fc receptor (FcRn) (Mandrup and Howard, 2021). Other strategies include to fusion to albumin targeting single-domain antibodies (sdAbs) (Tijink et al., 2008) or to albumin-binding peptides (Andersen et al., 2013). Multimerization is an additional approach to change the size and valency of antibody fragments to optimize its half-life and tumor penetration (Cuesta et al., 2010). In addition, multimerization allows the generation of multispecific molecules, which can improve their tissue specificity and endow antibodies with new functionalities. A number of strategies have been devised for multimersizing antibody fragments, such as domain-swapping approaches, linear gene fusions, and self-associating protein domains (Cuesta et al., 2010; Nunez-Prado et al., 2015). Among the latest strategies, it is worth mentioning the use of immunoglobulin (Ig)-derived oligomerization domains, mainly Ig heavy-chain constant (C_H) or Ig light-chain constant (C_L) domains, to generate formats such as the scFv-Fc antibody and the minibody (Brinkmann and Kontermann, 2017). A wide range of non-Ig protein oligomerization domains have also been used to generate multimeric antibodies, such as the bacterial barnase barstar pair, the Dock-and-Lock (DNL) method (based on the dimerization and docking domain of cAMP-dependent protein kinase A and the anchoring domain of an interactive A-kinase anchoring protein), and the coiled-coil peptides derived from human cartilage oligomeric matrix protein, among others (Brinkmann and Kontermann, 2017; Nunez-Prado et al., 2015). These modules have been successfully used to generate bivalent, trivalent, tetravalent and pentavalent antibodies (Brinkmann and Kontermann, 2017; Nunez-Prado et al., 2015).

3. The trimerbody concept and evolution

Trimerbodies are an emerging class of multimeric antibodies based on the unique structural and functional properties of trimerization domains of multiplexin collagens (Shoulders and Raines, 2009). The crystal structures for the trimerization domains of collagen XVIII and XV have been solved and both exhibit similar structural and biochemical properties (Boudko et al., 2009; Wirz et al., 2011). The collagen XVIII and XV trimerization domains are non-immunogenic, small-sized (less than 60 residues) and lack of disulphide bonds which enables fast folding. Trimers can form at picomolar or even lower concentrations and are extremely stable. Other advantages which make them attractive options for protein engineering are the high expression level and solubility, and the lack of N-glycosylation sites (Boudko et al., 2009; Wirz et al., 2011).

The trimerbody platform allows the development of multivalent (tri-, tetra- or hexavalent) and monospecific or multispecific (bi- or tri-specific) antibodies, with a format optimized for the desired pharmacokinetics and adapted to the pathological context (Cuesta et al., 2010). Multichain trimerbodies (MTs) are homotrimeric antibodies generated by fusing antigen-binding domains, such as scFvs or sdAbs, to the N- and/or C-terminus of a collagen-derived trimerization (TIE) domains, composed of the collagen XVIII or collagen XV trimerization regions flanked by peptide connectors (Fig. 1). First-generation MTs were engineered by fusing scFvs to the N-terminus or the C-terminus of the TIE domain of collagen XVIII (TIE^{XVIII}) (Sanchez-Arevalo, et al., 2006; Blanco-Toribio et al., 2013;) or XV (TIE^{XV}) (Cuesta et al., 2012). The resulting scFv-based monospecific antibodies (N-trimerbodies or C-trimerbodies; 110kDa) displayed enhanced antigen-binding capacity and stability. The trimerbody platform has also been used to generate scFv-based hexavalent molecules (N/C-trimerbodies; 190kDa) by fusing scFvs with the same or different specificity to both N- and C-terminus of a TIE^{XVIII} domain (Fig. 1). Monospecific or bispecific hexavalent N/C-trimerbodies are very efficient at recognizing antigen, either purified or expressed on the cell surface (Blanco-Toribio et al., 2013). Furthermore, N/C-trimerbodies containing scFvs at one terminus of the TIE^{XVIII} domain and sdAbs at the opposite terminus have also been successfully generated (asymmetric N/C-trimerbodies; 150kDa) (Compte et al., 2018, 2021; Mikkelsen et al., 2019) (Fig.1). Furthermore, fully functional asymmetric N/C-trimerbodies with a N-terminal tumor-specific scFv and a fungal ribotoxin in the C-terminus of the TIE^{XVIII} domain have also been generated (trimertoxins, 160kDa) (Lazaro-Gorines et al., 2019) (Fig. 1). Due to their multivalence, MTs demonstrated increased avidity, slow off-rates in vitro and improved tumor-targeting efficacy in vivo and anti-tumor effect in several mouse models of cancer (Compte et al., 2018, 2021). Additionally, MTs are efficiently produced as soluble functional

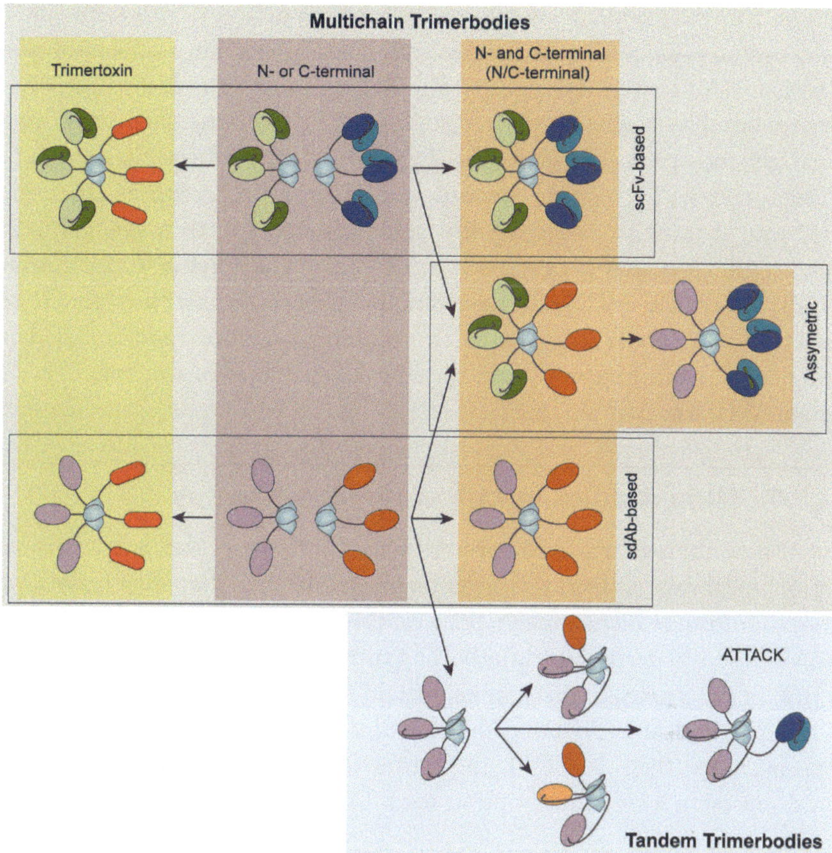

Fig. 1 Schematic diagrams showing the domain structure of trimerbody molecules. Multichain trimerbodies (gray box) are homotrimeric molecules generated by fusing single-chain variable fragments (scFvs) or single-domain antibodies (sdAbs), to the N- and/or C-terminus of a collagen-derived trimerization domain. Asymmetric trimerbodies contain scFvs at one terminus of the trimerization domain and sdAbs at the opposite terminus. Trimertoxins are asymmetric N/C-trimerbodies with a N-terminal tumor-specific scFv or sdAb and a C-terminal toxin. Tandem trimerbodies (blue box) are generated by fusing sdAbs to the N-terminus of the trimerization domains that are connected with additional peptide flexible linkers. The ATTACK (Asymmetric Tandem Trimerbody for T Cell Activation and Cancer Killing) is a T cell-recruiting trimerbody generated by fusing an anti-CD3 scFv to the C-terminus of a tumor-specific tandem trimerbody.

proteins in different expression systems including bacteria (Blanco-Toribio et al., 2015), yeast (Blanco-Toribio et al., 2014), insect and mammalian cells (Blanco-Toribio et al., 2013; Compte et al., 2018; Cuesta et al., 2009, 2012; Sanchez-Arevalo, et al., 2006).

The trimerbody platform allows the rational design of monospecific or multispecific molecules with defined stoichiometry within a single polypeptide chain (Alvarez-Cienfuegos et al., 2016). Tandem trimerbodies (TTs) are generated by connecting three sdAb-TIE^{XVIII} modules with two additional glycine-serine-based linkers (Fig. 1). TTs are efficiently secreted as soluble proteins by mammalian cells and are able to recognize their cognate antigen/s with high affinity and specificity (Alvarez-Cienfuegos et al., 2016). A recent evolution of the TT is an asymmetric T cell engager (TCE), called the ATTACK (asymmetric tandem trimerbody for T cell activation and cancer killing) which combines a tumor-specific TT with a single CD3-binding scFv fused to the C-terminus of the last TIE^{XVIII} domain (Harwood et al., 2017).

4. Trimerbodies in cancer immunotherapy

Since their emergence, trimerbodies have been used predominantly in preclinical cancer diagnosis and therapy. However, they may have application in multiple disease areas. For example, it has recently been generated a sdAb-based N-terminal trimerbody against the receptor-binding domain (RBD) of the SARS-CoV-2 Spike protein (Guttler et al., 2021). Compared to the parental anti-RBD sdAb, the trimerbody increased the blocking potency more than 300 fold, indicating that RBD-specific trimerbodies could represent a treatment option for COVID-19 (Guttler et al., 2021).

4.1 Immunomodulatory trimerbodies

Immune checkpoint receptors, both coinhibitory and costimulatory, are molecules expressed by cells of the immune system that regulate the activation and effector functions of T cells (Chen and Flies, 2013). The function of these receptors can be manipulated by the exogenous administration of mAbs to enhance T cell-mediated responses (Melero et al., 2015). The blockade of inhibitory checkpoint receptors, such as cytotoxic T lymphocyte-associated protein 4 (CTLA-4) and the programmed cell death-1 (PD-1, also known as CD279) receptor and its ligands [programmed death-ligand 1 (PD-L1, also known as B7-H1 or CD274), and PD-L2 (also known as B7-DC or CD273)], with antagonistic mAbs have shown outstanding efficacy in several types of cancer, with tolerable toxicity profiles; but their overall response rate remains around 30% (Baumeister et al., 2016). Agonistic mAbs targeting costimulatory checkpoint receptors, such as 4-1BB, [also known as tumor necrosis factor receptor superfamily member

9 (TNFRSF9) or CD137)], OX40 (also known as TNFRSF4 or CD134), CD40, glucocorticoid-induced TNFR-related protein [GITR, also known as TNFRSF18 or activation-inducible TNFR family receptor (AITR)], and inducible costimulatory molecule (ICOS, also known as CD278), are able to direct T cells toward an effector outcome and overcome anergy-inducing immunosuppressive signaling, thus providing strong rationale to be combined with immune checkpoint blockade (Melero et al., 2015). However, regardless significant effort, no such antibody has yet received regulatory approval. Among them, anti-4-1BB-agonistic mAbs have shown significant anti-tumor activity in preclinical models (Melero et al., 1997). However, off-tumor toxicities have been the mains obstacle for the clinical development of full-length anti-human 4-1BB-agonistic mAbs, and different studies suggest that the toxicity is mainly dependent on Fc–FcR interactions (Compte et al., 2018; Li and Ravetch, 2013; Xu et al., 2003).

Trimerbodies have two major advantages compared to IgG mAbs: they lack the Fc region involved in 4-1BB-mediated toxicity (Compte et al., 2018); and they are trimeric, as is the physiological 4-1BB ligand (4-1BBL). Recently we have reported MTs targeting tumor-associated antigens (TAAs), such as epidermal growth factor receptor [EGFR; (Compte et al., 2018)] or carcinoembryonic antigen [CEA; (Mikkelsen et al., 2019)], and 4-1BB in an agonistic manner. The bispecific hexavalent MT consists of three anti-4-1BB scFvs and three anti-TAA sdAbs organized around a TIE^{XVIII} module and has a 3:3 stoichiometry that is unique within the existing body of 4-1BB-specific proteins, including 4-1BBL fusion proteins (Fig. 2A) (Fellermeier et al., 2016).

In addition to an enhanced tumor-specific costimulation, the EGFR-targeted 4-1BB-agonistic trimerbody displayed an improved serum stability, as well as efficient and rapid targeting in EGFR-positive tumors. The trimerbody was able to maximize tumor-specific costimulation due to the combination of a higher tumor uptake and substantially faster clearance from blood circulation, ultimately giving an improved tumor to normal tissue ratio as compared to full-length 4-1BB-agonistic mAbs (Compte et al., 2018). In immunocompetent mice bearing murine colorectal carcinoma cells the EGFR-targeted 4-1BB-agonistic trimerbody exhibited anti-tumor activity similar to full-length anti-4-1BB-agonistic mAbs, and induce long-lasting response. Importantly, trimerbody treatment does not induce systemic cytokine production or liver toxicity associated with anti-4-1BB IgGs (Compte et al., 2018). The safety of the EGFR-targeted 4-1BB-agonistic trimerbody was further studied in an immunocompetent

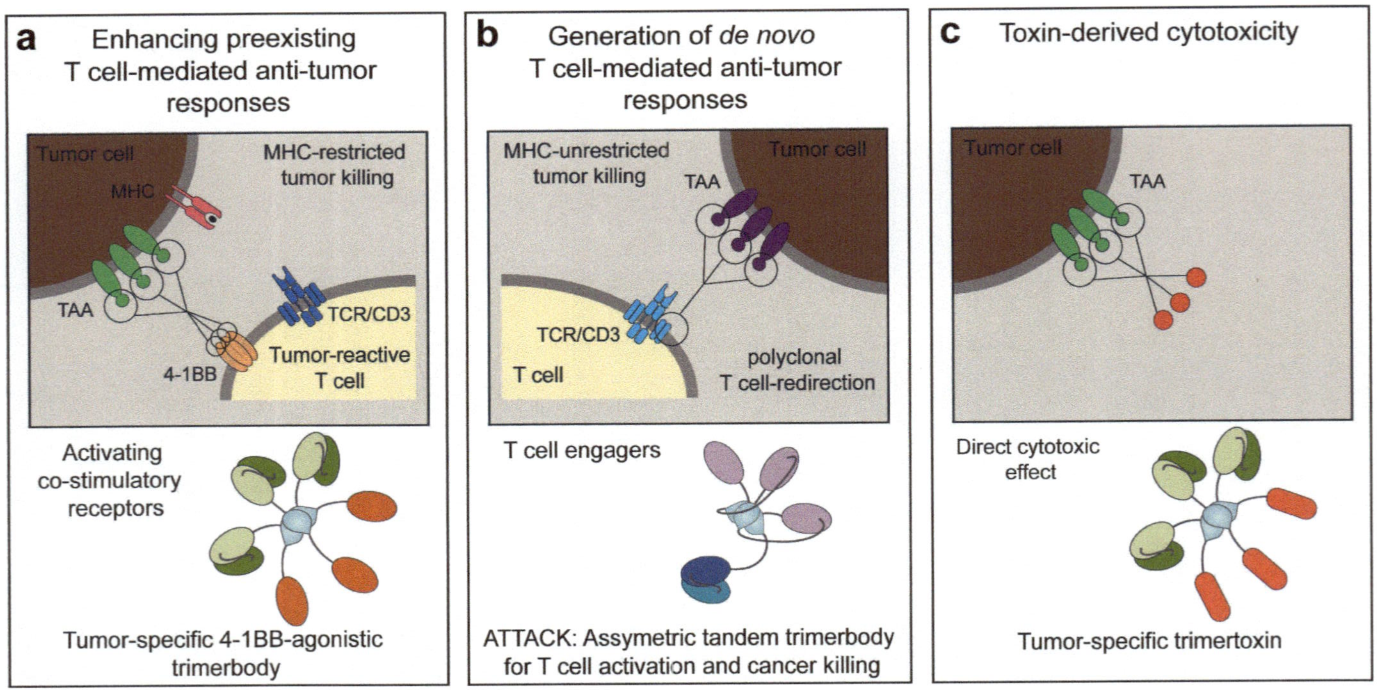

Fig. 2 Schematic diagram illustrating the mechanism of action of tumor-specific trimerbodies in cancer immunotherapy strategies. Trimerbodies for enhancement of antitumor activity (A) by activating costimulatory immune checkpoint receptors by TAA-targeted 4-1BB-agonistic multichain trimerbodies. Redirection of T cells toward extracellular TAA via pan-T cell-engaging trimerbodies (B). Direct cytotoxic effect by a TAA-specific multichain trimertoxin (C).

transgenic mouse model expressing human EGFR in the liver (Compte et al., 2020). Systemic administration of full-length anti-4-1BB-agonistic mAbs resulted in nonspecific immune stimulation and hepatotoxicity, whereas the trimerbody lacked these immune-related adverse effects, despite the fact that the anti-EGFR sdAb used recognize both human and mouse EGFR (Compte et al., 2020). These results validate EGFR as an effective target for the development of tumor-specific 4-1BB-mediated immunotherapy and support the development of an EGFR-targeted 4-1BB-agonistic trimerbody as a clinical candidate for treatment of solid tumors.

Therefore, a humanized EGFR-targeted 4-1BB-agonistic MT has been generated, which exhibits protein folding and structure nearly identical to that of the murine trimerbody (Compte et al., 2021). Despite the limitations of current mouse models for studying immunotherapy strategies against human tumors, such as the development of xenogeneic graft versus host disease (xGVHD) and insufficient engraftment of some human immune subpopulations (Morillon et al., 2020), we demonstrated that in human peripheral blood mononuclear cells-driven humanized immunoavatar mice bearing human colorectal cancer and triple-negative breast cancer (TNBC), the trimerbody monotherapy provided significant anti-tumor activity (Compte et al., 2021). This effect was confirmed in a humanized patient-derived xenograft model of human non-small cell lung cancer where treatment with the humanized EGFR-targeted 4-1BB-agonistic trimerbody showed a significant reduction in tumor growth, which was associated with a significant increase in the percentage of tumor infiltrating CD8+ cells and a substantial improvement of the CD8+ T cell/regulatory T cell ratio, in trimerbody-treated mice (Compte et al., 2021). Furthermore, in humanized mice bearing aggressive EGFR+ TNBC tumors expressing high levels of PD-L1, the combination of the humanized EGFR-targeted 4-1BB-agonistic MT with and immune checkpoint blocker further improved anti-tumor activity (Compte et al., 2021). In most of epithelial tumors more than 50% of patients display a moderate to strong expression of EGFR (Uhlen et al., 2015).

4.2 T cell-engaging trimerbodies

The earlier application of bispecific antibodies (bsAbs) in cancer immunotherapy was focused on redirecting T cells toward tumor cells. TCE bsAbs are designed to simultaneously binding to a specific TAA on the surface of cancer cells and one of the extracellular CD3 subunits (most commonly CD3ε) on the T cell surface (Blanco et al., 2021). By recognizing a user-defined TAA TCEs promote a major histocompatibility complex (MHC)-unrestricted tumor cell killing and are broadly applicable to virtually all cancer patients regardless of their MHC haplotype. This approach can bypass some of the evasion strategies used by tumors to escape recognition by the immune system, such as loss of MHC expression or defects in the antigen-presenting machinery (Ruiz-Cabello and Garrido, 1998). By adapting the TT format to the TAA/CD3 crosslinking approach, a novel class of Fc-free TCE trimerbody called ATTACK (asymmetric tandem trimerbody for T cell activation and cancer killing) has been generated (Fig. 2B) (Harwood et al., 2017). This tetravalent bispecific trimerbody combines trivalent binding to a cell surface TAA, and monovalent binding to CD3 with an intermediate size, achieving a balance between circulatory half-life extension and tumor penetration. Furthermore, the ATTACK positions the anti-TAA and anti-CD3 binding domains on opposite sides of the molecule, which facilitates efficient synapse formation between T lymphocytes and cancer cells (Harwood et al., 2017). We demonstrated that an anti-EGFR ATTACK (3 + 1) was more effective than an anti-EGFR light T cell engager (LiTE) antibody (1 + 1) (Molgaard et al., 2018) at inducing T cell activation and cytotoxicity against EGFR-positive cancer cells (Harwood et al., 2017). Under identical assay conditions, the ATTACK had a 15- to 20-fold higher potency than the LiTE and reached nearly 80% target cell lysis, compared to 60% reached by the LiTE (Harwood et al., 2017).

4.3 Trimerbody-based immunotoxins

Immunotoxins (ITX) combine the cell-killing ability of a toxin and the specificity of an antibody (Alewine et al., 2015). First- and second-generation ITX consisted of full-length mAb chemically coupled to a native toxin or its truncated form, respectively (Alewine et al., 2015). MAbs obtained from mouse hybridomas and toxins of plant or bacterial origin were highly immunogenic, precluding the administration of repeated doses (Alewine et al., 2015). Using recombinant DNA technology, the third generation of ITX was developed, overcoming past limitations and consisting primarily of an antibody fragment genetically fused to a catalytic toxin domain. In 2018, the FDA approved the first antibody-based ITX, moxetumomab pasudotox, for the treatment of hairy cell leukemia and another one, oportuzumab monatox was in review as of November 2021 for bladder cancer. Both of them consist of a scFv (anti-CD22 or anti-EpCAM, respectively) fused to a truncated version of *Pseudomonas* exotoxin A (Kreitman et al., 2021).

Although third generation ITX are amenable to multimerization strategies, few attempts of generating multimeric ITX have been reported. The production and characterization of a trimeric recombinant ITX targeting carcinoembryonic antigen (CEA) has recently been described. The CEA-specific "trimertoxin" was designed as asymmetric MT with a N-terminal anti-CEA MFE23 scFv and the fungal α-sarcin fused to the C-terminal end of the TIE^{XVIII} domain (Fig. 2C) (Lazaro-Gorines et al., 2019). The trimerbody-based ITX showed highly improved therapeutic effect in comparison to its monomeric counterpart, both in vitro and in vivo and. Indeed, the trimertoxin was significantly more effective in recognizing CEA-positive cancer cells and inhibiting protein synthesis ($IC_{50} = 6$ nM) than monomeric ITX ($IC_{50} = 60$ nM) (Lazaro-Gorines et al., 2019). Mice bearing human CEA-positive xenografts also demonstrated slower tumor growth rates and increased survival (Lazaro-Gorines et al., 2019). Overall, these results suggest that enhanced functional avidity of trimertoxins along with longer half-lives and improved tumor targeting capacity may represent a step forward to the development of next-generation ITX.

5. Trimerbodies in cancer diagnosis

Engineered antibodies for tumor imaging should have high specificity and affinity for the targeted TAA, rapid blood clearance rate and low accumulation in healthy tissues (Cuesta et al., 2010; Holliger and Hudson, 2005). It is well established that small bivalent antibodies, such as diabodies (55 kDa) might be ideal candidates for tumor imaging, as they show improved tumor-to-blood ratios than IgG antibodies (Adams et al., 2006; Viola-Villegas et al., 2014). Similarly, minibodies (80 kDa) resulting from the fusion of a scFv with the C_H3 domain of human IgG1 show fast blood clearance and rapid tumor targeting that reached maximum uptake at 12 h (Hu et al., 1996). The potential of the anti-CEA MFE23 scFv (Begent et al., 1996) for discrimination between tumor and normal tissues had been previously demonstrated using radioiodinated antibody (Chester et al., 2004). Iodine-123 (^{123}I)-labeled MFE23 scFv was found to be safe when administered intravenously and selectively localized gastrointestinal cancers, with a high tumor-to-blood ratio (Begent et al., 1996). A subsequent trial used ^{125}I- labeled MFE-23 scFv for radioimmunoguided surgery of primary or metastatic colorectal or pancreatic cancer patients (Mayer et al., 2000). Accurate tumor localization was also reported, with a sensitivity of 100% at 24h. Moreover, the importance of valency for efficient tumor targeting and retention was demonstrated using chemically cross-linked bivalent and trivalent A5B7 Fab' fragments in mice bearing human colorectal cancer xenografts (Casey et al., 1996). Increasing the number of binding domains (and therefore, avidity) in a single antibody molecule improved the amount of antibody and its residence time in tumors, showing favorable properties compared with the parental antibody.

In this line, we generated a near-infrared fluorescently labeled anti-CEA MFE23 scFv-based N-trimerbody and demonstrated that can specifically localize CEA-expressing xenografts (Cuesta et al., 2009). Moving forward toward a potential translation into the clinical setting, we have shown the tumor targeting capacity of the ^{124}I-labeled anti-CEA MFE23 N-trimerbody using positron emission tomography (PET) in a xenograft mouse model of human gastric adenocarcinoma (Rios et al., 2019). The trivalent MFE23 N-trimerbody showed higher tumor accumulation and tumor-to-muscle (T/M) ratios than the monovalent MFE23-scFv (Rios et al., 2019). Specific uptake was not detected with PET imaging in CEA-negative xenografts as indicated by low T/M ratios. Our data suggest that intermediate-sized trivalent trimerbodies could be promising candidates for targeted therapy and imaging of CEA-positive tumors.

6. Conclusions and future prospects

Bispecific antibodies are gaining momentum in many therapeutic areas, and more than 100 different formats have been described so far, including IgG and IgG-like molecules but also a plethora of small-sized antibodies. The main advantages of the trimerbody format are the small size, high expression level and solubility, the human origin of collagen XVIII and XV trimerization domains, and the robustness of the scaffold to generate multispecific molecules with a predictable and well-defined structure adapted to the pathological context. Preclinical efficacy and toxicity studies have demonstrated the potential of trimerbody-based molecules in T cell-mediated cancer immunotherapy strategies, either to enhance pre-existing responses or to generate de novo anti-tumor responses. In fact, the broad anti-tumor activity, synergy with immune checkpoint blockers and lack of toxicity demonstrated for the humanized tumor-specific 4-1BB-agonistic trimerbody has driven its clinical development and it is expected to launch a Phase I clinical trial in 2022. Furthermore, the significant advances in the development of engineered synthetic mRNA (Sanz and Alvarez-Vallina, 2021) could be instrumental in consolidating regulatory approvals for additional tumor-specific trimerbodies, In addition, the manufacturing problems associated with recombinant proteins will be avoided and production costs will be significantly reduced. We foresee a plethora of trimerbody designs for use in "avidifying" binding domains (Obeng et al., 2021) with increased therapeutic potential for cancer and non-cancer indications, such as infectious diseases including SARS-CoV-2, where trimerbodies could potentially bind and block the trimeric S protein to render the virus defective.

Acknowledgments

Financial support for this work was obtained from Spanish Ministry of Science and Innovation (SAF2017-89437-P and PID2020-117323RB-100 to LA-V), partially supported by the European Regional Development Fund (ERDF); the Carlos III Health Institute (ISCIII, PI19/00132 to LS, and DT20/00089 to LAV), partially supported by the ERDF; the Spanish Association Against Cancer (AECC 19084 to LA-V); and the CRIS Cancer Foundation (FCRIS-2018-0042 and FCRIS-2021-0090 to LA-V).

Competing interests

MC is current employee of Leadartis S.L. LS and LA-V are co-founders of Leadartis S.L.

References

Adams, G.P., Tai, M.S., McCartney, J.E., Marks, J.D., Stafford III, W.F., Houston, L.L., Huston, J.S., Weiner, L.M., 2006. Avidity-mediated enhancement of in vivo tumor targeting by single-chain Fv dimers. Clin. Cancer Res. 12, 1599–1605.

Alewine, C., Hassan, R., Pastan, I., 2015. Advances in anticancer immunotoxin therapy. Oncologist 20, 176–185.

Alvarez-Cienfuegos, A., Nunez-Prado, N., Compte, M., Cuesta, A.M., Blanco-Toribio, A., Harwood, S.L., Villate, M., Merino, N., Bonet, J., Navarro, R., Munoz-Briones, C., Sorensen, K.M., Molgaard, K., Oliva, B., Sanz, L., Blanco, F.J., Alvarez-Vallina, L., 2016. Intramolecular trimerization, a novel strategy for making multispecific antibodies with controlled orientation of the antigen binding domains. Sci. Rep. 6, 28643.

Andersen, J.T., Cameron, J., Plumridge, A., Evans, L., Sleep, D., Sandlie, I., 2013. Single-chain variable fragment albumin fusions bind the neonatal fc receptor (FcRn) in a species-dependent manner: implications for in vivo half-life evaluation of albumin fusion therapeutics. J. Biol. Chem. 288, 24277–24285.

Baumeister, S.H., Freeman, G.J., Dranoff, G., Sharpe, A.H., 2016. Coinhibitory pathways in immunotherapy for Cancer. Annu. Rev. Immunol. 34, 539–573.

Beckman, R.A., Weiner, L.M., Davis, H.M., 2007. Antibody constructs in cancer therapy: protein engineering strategies to improve exposure in solid tumors. Cancer 109, 170–179.

Begent, R.H., Verhaar, M.J., Chester, K.A., Casey, J.L., Green, A.J., Napier, M.P., Hope-Stone, L.D., Cushen, N., Keep, P.A., Johnson, C.J., Hawkins, R.E., Hilson, A.J., Robson, L., 1996. Clinical evidence of efficient tumor targeting based on single-chain Fv antibody selected from a combinatorial library. Nat. Med. 2, 979–984.

Blanco, B., Dominguez-Alonso, C., Alvarez-Vallina, L., 2021. Bispecific immunomodulatory antibodies for cancer immunotherapy. Clin. Cancer Res. 27, 5457–5464.

Blanco-Toribio, A., Sainz-Pastor, N., Alvarez-Cienfuegos, A., Merino, N., Cuesta, A.M., Sanchez-Martin, D., Bonet, J., Santos-Valle, P., Sanz, L., Oliva, B., Blanco, F.J., Alvarez-Vallina, L., 2013. Generation and characterization of monospecific and bispecific hexavalent trimerbodies. MAbs 5, 70–79.

Blanco-Toribio, A., Lacadena, J., Nunez-Prado, N., Alvarez-Cienfuegos, A., Villate, M., Compte, M., Sanz, L., Blanco, F.J., Alvarez-Vallina, L., 2014. Efficient production of single-chain fragment variable-based N-terminal trimerbodies in Pichia pastoris. Microb. Cell Fact. 13, 116.

Blanco-Toribio, A., Alvarez-Cienfuegos, A., Sainz-Pastor, N., Merino, N., Compte, M., Sanz, L., Blanco, F.J., Alvarez-Vallina, L., 2015. Bacterial secretion of soluble and functional trivalent scFv-based N-terminal trimerbodies. AMB. Express 5, 137.

Boudko, S.P., Sasaki, T., Engel, J., Lerch, T.F., Nix, J., Chapman, M.S., Bachinger, H.P., 2009. Crystal structure of human collagen XVIII trimerization domain: a novel collagen trimerization fold. J. Mol. Biol. 392, 787–802.

Brinkmann, U., Kontermann, R.E., 2017. The making of bispecific antibodies. MAbs. 9, 182–212.

Casey, J.L., King, D.J., Chaplin, L.C., Haines, A.M., Pedley, R.B., Mountain, A., Yarranton, G.T., Begent, R.H., 1996. Preparation, characterisation and tumour targeting of cross-linked divalent and trivalent anti-tumour Fab' fragments. Br. J. Cancer 74, 1397–1405.

Chen, L., Flies, D.B., 2013. Molecular mechanisms of T cell co-stimulation and co-inhibition. Nat. Rev. Immunol. 13, 227–242.

Chester, K., Pedley, B., Tolner, B., Violet, J., Mayer, A., Sharma, S., Boxer, G., Green, A., Nagl, S., Begent, R., 2004. Engineering antibodies for clinical applications in cancer. Tumour Biol. 25, 91–98.

Compte, M., Harwood, S.L., Munoz, I.G., Navarro, R., Zonca, M., Perez-Chacon, G., Erce-Llamazares, A., Merino, N., Tapia-Galisteo, A., Cuesta, A.M., Mikkelsen, K., Caleiras, E., Nunez-Prado, N., Aznar, M.A., Lykkemark, S., Martinez-Torrecuadrada, J., Melero, I., Blanco, F.J., Bernardino de la, S.J., Zapata, J.M., Sanz, L., Alvarez-Vallina, L., 2018. A tumor-targeted trimeric 4-1BB-agonistic antibody induces potent anti-tumor immunity without systemic toxicity. Nat. Commun. 9, 4809.

Compte, M., Harwood, S.L., Martinez-Torrecuadrada, J., Perez-Chacon, G., Gonzalez-Garcia, P., Tapia-Galisteo, A., Van Bergen En Henegouwen, P.M.P., Sanchez, A., Fabregat, I., Sanz, L., Zapata, J.M., Alvarez-Vallina, L., 2020. Case report: an EGFR-targeted 4-1BB-agonistic Trimerbody does not induce hepatotoxicity in transgenic mice with liver expression of human EGFR. Front. Immunol. 11, 614363.

Compte, M., Harwood, S.L., Erce-Llamazares, A., Tapia-Galisteo, A., Romero, E., Ferrer, I., Garrido-Martin, E.M., Enguita, A.B., Ochoa, M.C., Blanco, B., Oteo, M., Merino, N., Nehme-Alvarez, D., Hangiu, O., Dominguez-Alonso, C., Zonca, M., Ramirez-Fernandez, A., Blanco, F.J., Morcillo, M.A., Munoz, I.G., Melero, I., Rodriguez-Peralto, J.L., Paz-Ares, L., Sanz, L., Alvarez-Vallina, L., 2021. An fc-free EGFR-specific 4-1BB-agonistic trimerbody displays broad antitumor activity in humanized murine cancer models without toxicity. Clin. Cancer Res. 27, 3167–3177.

Cuesta, A.M., Sanchez-Martin, D., Sanz, L., Bonet, J., Compte, M., Kremer, L., Blanco, F.J., Oliva, B., Alvarez-Vallina, L., 2009. In vivo tumor targeting and imaging with engineered trivalent antibody fragments containing collagen-derived sequences. PLoS One. 4, e5381.

Cuesta, A.M., Sainz-Pastor, N., Bonet, J., Oliva, B., Alvarez-Vallina, L., 2010. Multivalent antibodies: when design surpasses evolution. Trends Biotechnol. 28, 355–362.
Cuesta, A.M., Sanchez-Martin, D., Blanco-Toribio, A., Villate, M., Enciso-Alvarez, K., varez-Cienfuegos, A., Sainz-Pastor, N., Sanz, L., Blanco, F.J., Alvarez-Vallina, L., 2012. Improved stability of multivalent antibodies containing the human collagen XV trimerization domain. MAbs 4, 226–232.
Fellermeier, S., Beha, N., Meyer, J.E., Ring, S., Bader, S., Kontermann, R.E., Muller, D., 2016. Advancing targeted co-stimulation with antibody-fusion proteins by introducing TNF superfamily members in a single-chain format. Onco. Targets. Ther. 5, e1238540.
Guttler, T., Aksu, M., Dickmanns, A., Stegmann, K.M., Gregor, K., Rees, R., Taxer, W., Rymarenko, O., Schunemann, J., Dienemann, C., Gunkel, P., Mussil, B., Krull, J., Teichmann, U., Gross, U., Cordes, V.C., Dobbelstein, M., Gorlich, D., 2021. Neutralization of SARS-CoV-2 by highly potent, hyperthermostable, and mutation-tolerant nanobodies. EMBO J. 40, e107985.
Harwood, S.L., Alvarez-Cienfuegos, A., Nunez-Prado, N., Compte, M., Hernandez-Perez, S., Merino, N., Bonet, J., Navarro, R., Van Bergen En Henegouwen, P.M.P., Lykkemark, S., Mikkelsen, K., Molgaard, K., Jabs, F., Sanz, L., Blanco, F.J., Roda-Navarro, P., Alvarez-Vallina, L., 2017. ATTACK, a novel bispecific T cell-recruiting antibody with trivalent EGFR binding and monovalent CD3 binding for cancer immunotherapy. Onco. Targets. Ther. 7, e1377874.
Holliger, P., Hudson, P.J., 2005. Engineered antibody fragments and the rise of single domains. Nat. Biotechnol. 23, 1126–1136.
Hu, S., Shively, L., Raubitschek, A., Sherman, M., Williams, L.E., Wong, J.Y., Shively, J.E., Wu, A.M., 1996. Minibody: a novel engineered anti-carcinoembryonic antigen antibody fragment (single-chain Fv-CH3) which exhibits rapid, high-level targeting of xenografts. Cancer Res. 56, 3055–3061.
Kontermann, R.E., 2009. Strategies to extend plasma half-lives of recombinant antibodies. BioDrugs 23, 93–109.
Kontermann, R.E., 2016. Half-life extended biotherapeutics. Expert Opin. Biol. Ther. 16, 903–915.
Kreitman, R.J., Dearden, C., Zinzani, P.L., Delgado, J., Robak, T., le Coutre, P.D., Gjertsen, B.T., Troussard, X., Roboz, G.J., Karlin, L., Gladstone, D.E., Kuptsova-Clarkson, N., Liu, S., Patel, P., Rotolo, F., Mitry, E., Pastan, I., Giles, F., 2021. Moxetumomab pasudotox in heavily pre-treated patients with relapsed/refractory hairy cell leukemia (HCL): long-term follow-up from the pivotal trial. J. Hematol. Oncol. 14, 35.
Lazaro-Gorines, R., Ruiz-de-la-Herran, J., Navarro, R., Sanz, L., Alvarez-Vallina, L., Martinez-Del-Pozo, A., Gavilanes, J.G., Lacadena, J., 2019. A novel carcinoembryonic antigen (CEA)-targeted trimeric immunotoxin shows significantly enhanced antitumor activity in human colorectal cancer xenografts. Sci. Rep. 9, 11680.
Li, F., Ravetch, J.V., 2013. Antitumor activities of agonistic anti-TNFR antibodies require differential FcgammaRIIB coengagement in vivo. Proc. Natl. Acad. Sci. U. S. A. 110, 19501–19506.
Mandrup, O.A., Howard, K.A., 2021. Bioengineered solutions to improve cancer immunotherapies. Ther. Deliv. 12, 339–341.
Mayer, A., Tsiompanou, E., O'Malley, D., Boxer, G.M., Bhatia, J., Flynn, A.A., Chester, K.A., Davidson, B.R., Lewis, A.A., Winslet, M.C., Dhillon, A.P., Hilson, A.J., Begent, R.H., 2000. Radioimmunoguided surgery in colorectal cancer using a genetically engineered anti-CEA single-chain Fv antibody. Clin. Cancer Res. 6, 1711–1719.
Melero, I., Shuford, W.W., Newby, S.A., Aruffo, A., Ledbetter, J.A., Hellstrom, K.E., Mittler, R.S., Chen, L., 1997. Monoclonal antibodies against the 4-1BB T-cell activation molecule eradicate established tumors. Nat. Med. 3, 682–685.

Melero, I., Berman, D.M., Aznar, M.A., Korman, A.J., Perez Gracia, J.L., Haanen, J., 2015. Evolving synergistic combinations of targeted immunotherapies to combat cancer. Nat. Rev. Cancer 15, 457–472.

Mikkelsen, K., Harwood, S.L., Compte, M., Merino, N., Molgaard, K., Lykkemark, S., varez-Mendez, A., Blanco, F.J., Alvarez-Vallina, L., 2019. Carcinoembryonic antigen (CEA)-specific 4-1BB-costimulation induced by CEA-targeted 4-1BB-agonistic trimerbodies. Front. Immunol. 10, 1791.

Molgaard, K., Harwood, S.L., Compte, M., Merino, N., Bonet, J., varez-Cienfuegos, A., Mikkelsen, K., Nunez-Prado, N., Alvarez-Mendez, A., Sanz, L., Blanco, F.J., Alvarez-Vallina, L., 2018. Bispecific light T-cell engagers for gene-based immunotherapy of epidermal growth factor receptor (EGFR)-positive malignancies. Cancer Immunol. Immunother. 67, 1251–1260.

Morillon, Y.M., Sabzevari, A., Schlom, J., Greiner, J.W., 2020. The development of next-generation PBMC humanized mice for preclinical investigation of cancer immunotherapeutic agents. Anticancer Res 40, 5329–5341.

Nimmerjahn, F., Ravetch, J.V., 2007. Fc-receptors as regulators of immunity. Adv. Immunol. 96, 179–204.

Nunez-Prado, N., Compte, M., Harwood, S., Alvarez-Mendez, A., Lykkemark, S., Sanz, L., Alvarez-Vallina, L., 2015. The coming of age of engineered multivalent antibodies. Drug Discov. Today 20, 588–594.

Obeng, E.M., Dzuvor, C.K.O., Danquah, M.K., 2021. Anti-SARS-CoV-1 and -2 nanobody engineering towards avidity-inspired therapeutics. Nano Today 42, 101350.

Reichert, J.M., Dhimolea, E., 2012. The future of antibodies as cancer drugs. Drug Discov. Today 17, 954–963.

Rios, X., Compte, M., Gomez-Vallejo, V., Cossio, U., Baz, Z., Morcillo, M.A., Ramos-Cabrer, P., Alvarez-Vallina, L., Llop, J., 2019. Immuno-PET imaging and pharmacokinetics of an anti-CEA scFv-based Trimerbody and its monomeric counterpart in human gastric carcinoma-bearing mice. Mol. Pharm. 16, 1025–1035.

Ruiz-Cabello, F., Garrido, F., 1998. HLA and cancer: from research to clinical impact. Immunol. Today 19, 539–542.

Sanchez-Arevalo, L.V., Cuesta, A.M., Sanz, L., Compte, M., Garcia, P., Prieto, J., Blanco, F.J., Alvarez-Vallina, L., 2006. Enhanced antiangiogenic therapy with antibody-collagen XVIII NC1 domain fusion proteins engineered to exploit matrix remodeling events. Int. J. Cancer 119, 455–462.

Sanz, L., Alvarez-Vallina, L., 2021. Engineered mRNA and the rise of next-generation antibodies. Antibodies (Basel) 10.

Sanz, L., Cuesta, A.M., Compte, M., Alvarez-Vallina, L., 2005. Antibody engineering: facing new challenges in cancer therapy. Acta Pharmacol. Sin. 26, 641–648.

Scott, A.M., Wolchok, J.D., Old, L.J., 2012. Antibody therapy of cancer. Nat. Rev. Cancer 12, 278–287.

Shoulders, M.D., Raines, R.T., 2009. Collagen structure and stability. Annu. Rev. Biochem. 78, 929–958.

Therapeutic Monoclonal Antibodies Approved or in Review in the EU or US. https://www.antibodysociety.org/resources/approved-antibodies/. 2021. Ref Type: Electronic Citation.

Tijink, B.M., Laeremans, T., Budde, M., Stigter-van, W.M., Dreier, T., de Haard, H.J., Leemans, C.R., van Dongen, G.A.M.S., 2008. Improved tumor targeting of anti-epidermal growth factor receptor nanobodies through albumin binding: taking advantage of modular nanobody technology. Mol. Cancer Ther. 7, 2288–2297.

Uhlen, M., Fagerberg, L., Hallstrom, B.M., Lindskog, C., Oksvold, P., Mardinoglu, A., Sivertsson, A., Kampf, C., Sjostedt, E., Asplund, A., Olsson, I., Edlund, K., Lundberg, E., Navani, S., Szigyarto, C.A., Odeberg, J., Djureinovic, D., Takanen, J.O., Hober, S., Alm, T., Edqvist, P.H., Berling, H., Tegel, H., Mulder, J., Rockberg, J., Nilsson, P., Schwenk, J.M., Hamsten, M., von, F.K., Forsberg, M., Persson, L., Johansson, F., Zwahlen, M., von, H.G., Nielsen, J., Ponten, F., 2015. Proteomics. Tissue-based map of the human proteome. Science 347, 1260419.

Viola-Villegas, N.T., Sevak, K.K., Carlin, S.D., Doran, M.G., Evans, H.W., Bartlett, D.W., Wu, A.M., Lewis, J.S., 2014. Noninvasive imaging of PSMA in prostate tumors with (89)Zr-labeled huJ591 engineered antibody fragments: the faster alternatives. Mol. Pharm. 11, 3965–3973.

Wirz, J.A., Boudko, S.P., Lerch, T.F., Chapman, M.S., Bachinger, H.P., 2011. Crystal structure of the human collagen XV trimerization domain: a potent trimerizing unit common to multiplexin collagens. Matrix Biol. 30, 9–15.

Xenaki, K.T., Oliveira, S., van Bergen En Henegouwen, P.M.P., 2017. Antibody or antibody fragments: implications for molecular imaging and targeted therapy of solid tumors. Front. Immunol. 8, 1287.

Xu, Y., Szalai, A.J., Zhou, T., Zinn, K.R., Chaudhuri, T.R., Li, X., Koopman, W.J., Kimberly, R.P., 2003. Fc gamma Rs modulate cytotoxicity of anti-Fas antibodies: implications for agonistic antibody-based therapeutics. J. Immunol. 171, 562–568.

CHAPTER FOUR

Anticalin®-based therapeutics: Expanding new frontiers in drug development

Aizea Morales-Kastresana, Martin Siegemund, Stefan Haak, Janet Peper-Gabriel, Vanessa Neiens, and Christine Rothe*

Pieris Pharmaceuticals GmbH, Hallbergmoos, Germany

*Corresponding author: e-mail address: rothe@pieris.com

Contents

Abstract

Anticalin proteins are a novel class of clinical-stage biopharmaceuticals with high potential in various disease areas. Anticalin proteins, derived from extracellular human lipocalins are single-chain proteins, with a highly stable structure that can be engineered to bind with high specificity and potency to targets of therapeutic relevance. The small size and stable structure support their development as inhalable biologics in the field of respiratory diseases as already demonstrated for PRS-060/AZD1402, an Anticalin protein currently undergoing clinical development for the treatment of asthma. Anticalin proteins provide formatting flexibility which allows fusion with the same or other Anticalin proteins, or with other biologics to generate multivalent, multiparatopic or multispecific fusion proteins. The fusion of Anticalin proteins to antibodies allows the generation of potent therapeutic proteins with new modes of action, such as antibody-Anticalin bispecific proteins with tumor-localized activity. Cinrebafusp alfa and PRS-344/S095012 antibody-Anticalin bispecific proteins were designed to reduce potential systemic toxicity by localizing the activity to the tumor,

International Review of Cell and Molecular Biology, Volume 369
ISSN 1937-6448
https://doi.org/10.1016/bs.ircmb.2022.03.009

and are currently in clinical development in immuno-oncology. Furthermore, the ease in generating bi- and multispecifics as well as the small and stable structure prompted the investigation of Anticalin proteins for the CAR T space, opening additional potential treatment options based on Anticalin protein therapies.

1. Introduction

Biopharmaceuticals and especially antibodies as highly specific immune-modulators, have revolutionized therapies in the fields of chronic inflammatory disease as well as cancer. The surge of antibody-based therapeutic approaches and the great amount of bioengineering performed to optimize their developability and therapeutic performance have proven their benefits as a therapeutic modality. In fact, 100 antibody treatments were approved until April 2021 by the FDA, mostly in the field of cancer, but also in 12 other therapeutic areas, including infectious and autoimmune diseases (Mullard, 2021).

Despite the great medical advancement based on antibody therapies, it is evident that in order to modulate the immune system effectively and safely, deep knowledge of the targeted pathways, high therapeutic precision, and, potentially, multitargeting properties within the biological agent are required. Forceful stimulatory triggers may be needed to overcome the immunosuppressive environment in an advanced solid tumor, and equally forceful immune inhibition may be required in inflammatory diseases to allow recovery of the affected tissue. Systemic modulation of the immune system, however, can have detrimental side effects and hence, must be confined in space and time. At the site of the tumor lesion, immune cells act in a highly coordinated fashion embedded in a network of signals with inbuilt feedback mechanisms and close interaction with the local tissue. To be curative in a chronic disease setting, the therapeutic intervention must account for the level of complexity of the immune system. Technically, this requires therapeutic modalities set up for multispecificity, providing control over valency, geometry and target-to-target distance. To implement this in a purely antibody-based design, heavy bioengineering is required, with potential consequences on developability, manufacturability and tolerability, evincing limitations of antibody therapies.

Anticalin proteins extend the spectrum of drug format and application options and specifically address some of the shortcomings of monoclonal antibodies, as they offer a high degree of formatting flexibility, stability and potency in a compact molecular format.

Anticalin proteins originate from the human secreted lipocalin protein family and were first established by Prof. Arne Skerra and colleagues at the Technical University of Munich, which resulted in the foundation of the biotech company Pieris Pharmaceuticals (Deuschle et al., 2021; Richter et al., 2014). Lipocalins represent a family of extracellular proteins abundant in all higher organisms with the function to store, transport or sequester, ligands ranging from small molecules to proteins (Skerra, 2000).

At Pieris, the Anticalin protein technology was further developed into a state-of-the-art therapeutics platform, with 3 programs currently in clinical development in the fields of cancer immunotherapy and respiratory disease (NCT04643158, NCT03330561, NCT03650348 and EudraCT2019-003456-36). Beyond that, Pieris runs multiple preclinical projects within an extended indication space, including infectious disease and ophthalmology (Rothe and Skerra, 2018).

Multispecific therapeutic approaches are at the heart of the Anticalin protein technology, as they can function as universal target-binding modules applicable on their own or in more complex Anticalin protein based bi- and multispecific fusion formats. Anticalin protein's properties make them excellent building blocks for localized functionality either via design-driven tissue targeting, or local application.

In the following sections we provide details on the technology platform and lay out use cases exemplifying the excellent properties of Anticalin proteins for a diverse set of therapeutic approaches in different disease areas, with a special focus on cancer and respiratory diseases.

2. Anticalin protein technology

Lipocalins are small, monomeric proteins with a molecular mass of about 20 kDa (Skerra, 2008). They have a low sequence homology but share a common stable 8-stranded β-barrel structure with an α-helix attached to the central barrel at the C-terminal side. One side of the barrel is closed by short loops and a dense packing of side chains, while the opposite side of the barrel is open with four flexible protruding loops, reminiscent of CDR loops in antibodies, varying in length and structure between the members of the lipocalin family (Fig. 1) (Schlehuber and Skerra, 2005).

To date 15 lipocalins have been characterized in humans. Based on biochemical and structural characterization, two of them, the tear lipocalin (Tlc), also called lipocalin 1 (Lcn 1) and the neutrophil gelatinase associated lipocalin (NGAL), also called lipocalin 2 or siderocalin, were chosen as

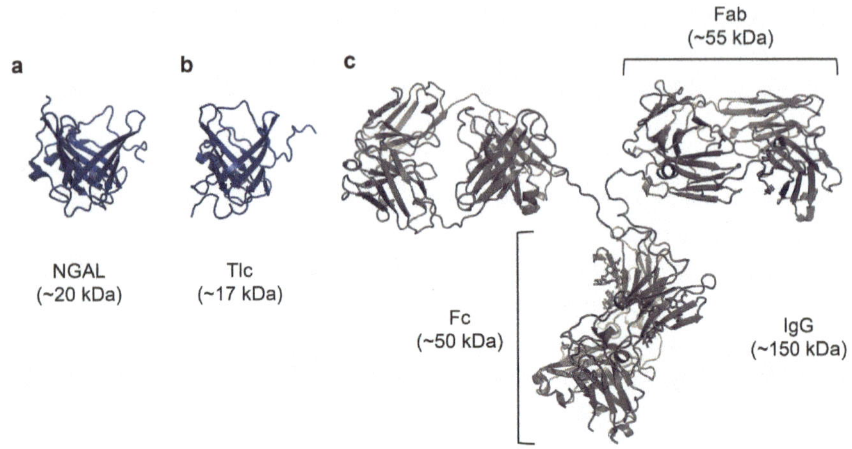

Fig. 1 Three-dimensional structure of neutrophil-gelatinase associated lipocalin (NGAL, PDB 1DFV) (A) and tear lipocalin (Tlc, PDB5T43) (B) shown in scale to an antibody in the IgG format (PDB 1IGT) (C).

scaffolds to build the Anticalin protein libraries (Breustedt et al., 2005, 2009; Goetz et al., 2002; Rothe and Skerra, 2018; Schiefner and Skerra, 2015). Tear lipocalin is the major protein in the tear fluid and seems to play a role in preserving the integrity of the tear film (Breustedt et al., 2005). NGAL and also Tlc play a role in the innate immune system by scavenging bacterial siderophores and thus inhibiting bacterial growth (Flo et al., 2004). Both lipocalins could be engineered to abrogate binding capacity to the natural ligands.

NGAL- and Tlc- based Anticalin protein libraries, with various designs and each library including about 10^{10} unique members, were generated using rational design and protein engineering methods to diversify amino acid positions in the flexible loop regions, the inner side of the pocket, or the α-helical region. Those Anticalin protein libraries are used for standard phage and yeast display technologies. Automated screening procedures to select high affinity binders against a variety of different targets have been established (Rothe and Skerra, 2018). Crystal structures have shown that Anticalin proteins provide a large target binding surface comparable to, and occasionally even larger than, conventional antibodies (Schonfeld et al., 2009).

A defined characterization workflow was established to screen the Anticalin proteins not only for target binding and functionality, but also for developability as it has become evident that drug-like properties, such as the absence of off-target binding, a reduced risk of immunogenicity, storage stability, and manufacturing feasibility, are of critical importance for successful drug development (Jain et al., 2017). The established workflow furthermore enables the assessment of the robustness of drug candidates upon mechanical stress, which is a prerequisite for the development of Anticalin proteins in respiratory diseases where the potential for inhaled delivery is a key advantage of Anticalin proteins over many other biologics (Skerra, 2008).

Anticalin proteins have proven suitable for prokaryotic expression in *E. coli* (Gille et al., 2008; Hohlbaum et al., 2018) or *Corynebacterium glutamicum* (CORYNEX®, Ajinomoto Co., Inc. (Tokyo, Japan)) (Masuda et al., 2018) as well as for mammalian expression in HEK or CHO cells (Hinner et al., 2019). Furthermore, manufacturing processes have successfully been scaled up to commercial scale to produce Anticalin proteins as well as Anticalin protein based bi- and multispecific fusion proteins for clinical application (Hinner et al., 2019).

3. Format opportunities for Anticalin proteins

Anticalin proteins can be developed as small, monomeric "Monocalins" for local administration in the lung as shown for PRS-060/AZD1402, an inhaled Anticalin protein targeting IL-4 receptor alpha (IL-4Ra) to treat moderate-to-severe asthma currently in a phase 2 clinical trial (NCT04643158). For other therapeutic modalities, Anticalin proteins can be fused flexibly at their N- or C-terminus to generate tandem or multidomain proteins, respectively, with other molecules to achieve higher degrees of valency, specificity, or half-life extension in plasma. Anticalin proteins can be genetically fused with each other to build dimers (Duocalin® proteins) (Richter et al., 2014), trimers (Tricalins) or tetramers (Tetracalins), facilitating bi- or multivalent, and bi- or multispecific, binding (Fig. 2). Aside from formats exclusively composed of Anticalin proteins, other molecules like IgG or Fc have been used as fusion partners with Anticalin proteins providing additional target specificity, avidity, extended pharmacokinetic profile, or effector functions. The fusion of Anticalin proteins to existing antibodies allows the generation of therapeutic proteins with new modes of action, as demonstrated for the HER2/4-1BB (CD137) antibody-Anticalin bispecific protein, cinrebafusp alfa (also known as PRS-343) (Hinner et al., 2019). The formatting flexibility of Anticalin proteins, which allows their fusion to either the N- or the C-terminus of

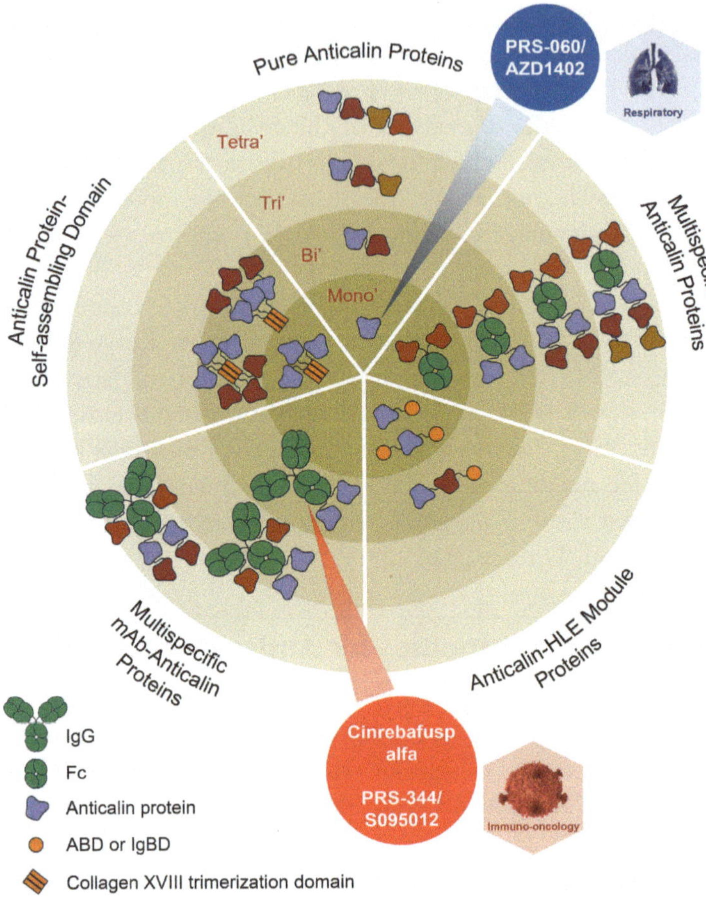

Fig. 2 Format options for Anticalin proteins. Anticalin therapeutics can exert their function as monomeric and therefore monospecific small biologics, e.g., in the background of inhaled therapeutics. Multivalent as well as multispecific formats can either be purely Anticalin protein-based or can be created by genetic arrangement of different molecular building blocks including Anticalin proteins, monoclonal antibodies (mAbs) or IgG Fc moieties. Fusions of Anticalin proteins with small plasma half-life extension modules such as albumin binding domain (ABD) or immunoglobulin binding domain (IgBD) enable in contrast to IgG or Fc a monovalent configuration of antigen binding sites. Fusions of Anticalin proteins with self-assembling oligomerization domains represent another option to achieve higher degrees of valency or specificity.

the heavy or light chain of antibodies, offers the possibility to generate bispecific compounds with different geometries and spacing between IgG paratopes and Anticalin protein binding sites, which may influence and fine-tune activity (Hinner et al., 2019). Beyond that, Anticalin fusions with half-life extension modules like the albumin binding domain (ABD) or the immunoglobulin binding domain (IgBD) have been proven among others for pharmacokinetic engineering (Siegemund et al., 2021). Anticalin protein fusions with the self-assembling human collagen XVIII trimerization (TIE^{XVIII}) domain, previously described for making multispecific antibodies (trimerbodies) (Alvarez-Cienfuegos et al., 2016), have recently been added to the list of format opportunities opening additional therapeutic applications as outlined below.

4. Plasma half-life engineering of Anticalin proteins

The development of Anticalin proteins for diagnostic and therapeutic purposes is accompanied by the development of strategies for tailored, application-oriented, plasma half-life. Due to their small size, single Anticalin proteins are rapidly cleared from circulation by renal filtration which can prevent sufficient biodistribution into the target tissue. In preclinical rodent models, Monocalins have shown terminal plasma half-lives mostly below 1 h (Gille et al., 2016; Masuda et al., 2018; Siegemund et al., 2021). Multiple strategies are available to improve the pharmacokinetic profile of Anticalin proteins without affecting their capacity for target binding. Plasma half-life extension (HLE) of Anticalin proteins has been achieved by increasing the hydrodynamic radius or by utilizing FcRn-recycled proteins like albumin, IgG and the IgG Fc moiety as partners for genetic fusion or non-covalent binding. By using these strategies, a range of relevant pharmacokinetic profiles reaching from intermediate to IgG-like half-life has become available.

PEGylation and PASylation® are established technologies for increasing the hydrodynamic volume of the linked protein to prevent a rapid excretion from the bloodstream via the kidneys. The VEGF-A- neutralizing Anticalin protein, PRS-050, developed for treatment of neovascular diseases like cancer (Mross et al., 2013) and wet age-related macular degeneration, was conjugated in a site-directed manner with different entities of PEG-maleimide, which extended the terminal plasma half-life ($t_{1/2}$) up to 28 h (Gille et al., 2016). Analogously, coupling to a 30 kDa PEG variant extended the $t_{1/2}$ of the hepcidin-antagonist Anticalin protein, PRS-080, for treatment of iron homeostasis disorders to 43 h in cynomolgus monkeys (Hohlbaum et al., 2018) and up to 80 h in healthy humans (Renders et al., 2019). Aiming for moderate plasma HLE that is beneficial in imaging applications *in vivo*, a ^{89}Zr-labeled Anticalin protein directed against the ectodomain of the human tumor-marker CD98hc was genetically fused with a structurally disordered polypeptide comprising in total 200 proline, alanine and serine residues (PAS200). This pharmacokinetic fine-tuning by PASylation® was used as a strategy to boost protein tracer accumulation and tumor contrast (Deuschle et al., 2020).

The utilization of the FcRn-recycled serum proteins, albumin and IgG, is besides the increase of the hydrodynamic radius another effective strategy for plasma HLE of Anticalin proteins. Non-covalent binding to those serum proteins by streptococcal albumin or immunoglobulin binding domains (ABD, IgBD) has been adapted for pharmacokinetic tuning, in particular when aiming for intermediate plasma half-lives. Masuda et al., fused an anti-PCSK9 Anticalin protein with a deimmunized high-affinity ABD, which extended the $t_{1/2}$ of the molecule from ~1 h to ~24 h in rats (Masuda et al., 2018). Different affinities of ABD and IgBD to their respective target serum proteins have been postulated to correlate with HLE of exemplary anti-OX40 Anticalin proteins and provide a "toolbox" for tailored modulation of pharmacokinetics, e.g., by use of IgBDs in a modular manner to exert longer plasma half-life through avidity effects (Siegemund et al., 2021). Such monomeric modules also represent a HLE option for monovalent Anticalin protein antagonists directed against receptors that would potentially be activated by bivalent IgG/Fc-based protein formats.

Fusions of Anticalin proteins with monoclonal antibodies (mAbs) represent the most common therapeutic formats and combine a long half-life with bi- or even multispecificity. For the molecule cinrebafusp alfa, composed of two anti-4-1BB Anticalin proteins fused to each one of the heavy chain C-termini of an anti-HER2 antibody, terminal elimination half-lives of >14 days in wild-type mice and approximately 4 days in cynomolgus monkeys were observed, mirroring the behavior of conventional mAbs in preclinical models (Hinner et al., 2019). Apart from IgG, fusion to Fc was shown to extend the $t_{1/2}$ in the case of an anti-OX40 Anticalin protein up to 6 days in human serum albumin (hSA)/human FcRn double humanized mice (Siegemund et al., 2021).

5. Anticalin protein based multispecifics for potential treatment of infectious diseases

Anticalin based multispecifics can also be applied to generate potential therapeutics to fight infectious diseases. Bacteria need iron for growth and virulence and use low-molecular-weight chelating agents, called siderophores, for iron sequestration (Braud et al., 2009). *Pseudomonas aeruginosa* (*P. aeruginosa*) is a human opportunistic pathogen and one of the major pathogens in nosocomial infections affecting immune compromised patients and patients suffering from tissue injuries or cystic fibrosis. *P. aeruginosa* expresses many different siderophores that can be classified as pyoverdines (Pvd) and pyochelin (pch) (Braud et al., 2009).

Four different Anticalin proteins were selected from NGAL based Anticalin protein libraries with specificity for Pvd or pch proteins (Dauner and Skerra, 2020). Anticalin multimers were generated by genetic fusion of the four Anticalin proteins, to bind and block all Pvd and pch siderophores from the different clinically relevant *P. aeruginosa* strains with one single therapeutic molecule (Corvey et al., 2016). For systemic application, the Tetracalin was genetically fused to an Fc domain for half-life extension. The Tetracalin was able to inhibit bacterial growth both *in vitro* as well as in cystic fibrosis mouse models and chronic lung infection rat models *in vivo*. Furthermore, the drug candidate showed activity on most of the cystic fibrosis clinical isolates and significantly reduced the bacterial load (Pieris Pharmaceuticals Inc., 2015 & unpublished data).

6. Antibody-Anticalin fusion formats for cancer immunotherapy

Immuno-oncology is a fast-growing field propelled by the success of Ipilimumab, the first antibody therapeutic targeting an immunomodulatory receptor on T cells, approved more than a decade ago (Hodi et al., 2010). Since then, many antibodies targeting key immunomodulatory receptors have received approval in various cancer indications. Nonetheless, there is an urgent medical need for the treatment of patients that do not respond to currently approved immunotherapies, as well as to improve the safety profile of existing therapies (Waldman et al., 2020).

One of the caveats of immunotherapies is the risk of activity-linked systemic toxicities (Das and Johnson, 2019; Perez-Ruiz et al., 2019). Strategies that aim to minimize the risk of systemic and broad activation of the immune system are therefore extremely important. Some explored strategies include optimizing the route of administration, hitting the right target on the right cells, applying ex vivo expansion of specific immune cell subsets, or on/off systems that fine tune the immune response in the event of toxicities (Arnaud et al., 2021; Jan et al., 2021; Mitchell et al., 2021; Petersen et al., 2018; Zhao et al., 2019). Another approach to avoid systemic activity is to localize the immune activation to the site of the tumor. Bispecific compounds represent a major advancement in the localized immune activation approach, and currently, several bispecific compounds are in clinical and preclinical development (Blanco et al., 2021; Ma et al., 2021). Several formats are possible, but generally these bispecific compounds contain two "arms": one arm serves as the anchor that targets the bispecific to the tumor microenvironment, while the other arm induces the immune-activating function selectively in the tumor tissue. Cinrebafusp alfa, an antibody-Anticalin fusion protein addressing the tumor target HER2 and the T cell target 4-1BB was designed to localize the immune activation to the site of the tumor. 4-1BB (CD137) is a T cell co-receptor mainly expressed on the membrane of activated and exhausted T cells. The 4-1BB pathway is activated by clustering several 4-1BB molecules, either via its natural trimeric ligand 4-1BBL, or via exogenous molecules like antibodies, antibody derivatives, aptamers, multimeric proteins and peptides, or bispecifics (Claus et al., 2019; Compte et al., 2021; Hinner et al., 2019; Hurov et al., 2021; McNamara et al., 2008; Melero et al., 1998; Schrand et al., 2014). Preclinical models support the development of 4-1BB targeting molecules to drive effective antitumoral responses (Melero et al., 1998, 2008). However, clinical development of 4-1BB agonistic antibodies has been hindered by either lack of activity or systemic toxicity (Fisher et al., 2012; Segal et al., 2017). To overcome this limitation, Hinner and colleagues developed cinrebafusp alfa, a potent first-in-class HER2/4-1BB antibody-Anticalin bispecific by fusing a 4-1BB-targeting Anticalin protein to an anti-HER2 antibody, and thus directing the activity of the anti-4-1BB Anticalin protein to the HER2+ tumor microenvironment. Tumor-localized T cell activation bears the potential to reduce the risk for systemic toxicity (Hinner et al., 2019). This bispecific was tested in patients with a variety of solid tumors that overexpress HER2 and demonstrated an acceptable safety profile and anti-tumor activity in a heavily pre-treated patient population both as a single agent and in combination with atezolizumab (NCT03330561 and NCT03650348, respectively). The administration of cinrebafusp alfa led to a dose-dependent activation of the immune system demonstrated by an increase of CD8+T cells in the tumor microenvironment. The efficacy of cinrebafusp alfa was observed in tumors already infiltrated by T cells ("hot tumor") and in non-immunogenic "cold" tumors not yet infiltrated by T cells, a population considered non-responsive to conventional checkpoint inhibition. Furthermore, the activity of 4-1BB targeting Anticalin proteins was confirmed by the elevation of soluble 4-1BB in serum, a downstream biomarker of 4-1BB target engagement and pathway activation (Piha-Paul et al., 2021). Through this mode of action, cinrebafusp alfa demonstrated a durable anti-tumor activity and the encouraging results led to continuing the clinical development of cinrebafusp alfa in phase 2 studies.

To extend the success of 4-1BB-targeting bispecific therapeutics beyond HER2+ tumors, the 4-1BB Anticalin protein was fused to a PD-L1 targeting antibody. PD-L1 is a pan-cancer checkpoint receptor that is expressed on a variety of tumor cells, stromal cells and myeloid cells infiltrating the tumor microenvironment (Oh and Bang, 2020; Yi et al., 2021). PD-L1 is a ligand for PD-1, a negative regulator of T cell activation and function (Zou et al., 2016). In the clinical setting, blockade of PD-1: PD-L1 interaction augments the T cell activity and anti-tumoral immune responses, ultimately extending patient survival. However, a significant number of patients either do not initially respond to anti-PD-(L)1 therapies or develop resistance mechanisms (Jenkins et al., 2018). PRS-344/S095012 is a PD-L1/4-1BB-targeting antibody-Anticalin bispecific, which acts via a triple mechanism: (i) blocking the PD-1:PD-L1 pathway; (ii) clustering 4-1BB and thus activating the 4-1BB pathway on activated and exhausted T cells in the tumor; and (iii) avoiding 4-1BB activation in the peripheral tissues that are lower in PD-L1 expression. This molecule recapitulates cinrebafusp alfa's preclinical efficacy and is currently in clinical testing in patients with a variety of solid tumors (EudraCT Number: 2019-003456-36) (Peper-Gabriel et al., 2022).

7. Armoring cancer immunotherapies with Anticalin protein-based payloads

Another field of rapid innovation and great therapeutic potential in cancer therapy is adoptive cell therapy, in particular treating patients with chimeric antigen receptor (CAR) T cells. To date, CAR T cell therapy has achieved remarkable response rates in patients with B cell lymphoma (Schuster et al., 2017) and acute lymphoblastic leukemia (ALL) (Park et al., 2018). Beyond hematological cancers, the efficacy of CAR T cells, so far, has been limited (reviewed in Jackson et al., 2016). To extend the clinical success of CAR T cells to solid tumors, advanced designs of CAR T cells are currently being explored. One such approach is to generate "armored CAR T cells" which are further genetically engineered to co-express immune-modulating proteins to enhance CAR T cell efficacy (Jackson et al., 2016; Yeku et al., 2017; Rafiq et al., 2018).

Anticalin proteins provide high functional density in a small coding size, which qualifies them for the CAR T cell approach not only as binding moiety but also as ideal candidates for secreted immune modulatory cargo proteins in armored CAR T cell therapy. The approach of Anticalin protein armored CAR T cells results in tumor-localized accumulation of the immunomodulatory Anticalin proteins, specifically promoting anti-tumor immunity, whereas systemic exposure and associated potential toxicity is limited due to the short intrinsic half-life of Anticalin proteins. A further potential advantage of utilizing Anticalin proteins and their high format flexibility could be the generation of multispecifics containing a tumor targeting moiety that could maximize the Anticalin protein accumulation in the tumor microenvironment.

4-1BB signaling elements in the cytoplasmic domain of CAR molecules have been previously described to drive durable and effective clinical responses (Maude et al., 2014). Armoring CAR T cells with secreted 4-1BB activating Anticalin protein multimers could further support the persistence of CAR and endogenous T cells in both an autocrine and paracrine manner, respectively. This hypothesis is supported by preclinical data demonstrating increased CAR T cell efficacy in combination with systemic anti-4-1BB antibody therapy (Mardiana et al., 2017). To allow efficient 4-1BB receptor clustering, a hexavalent anti-4-1BB engager was generated by fusing two 4-1BB targeting Anticalin proteins to the human collagen XVIII trimerization (TIE^{XVIII}) domain, separated by a serine-glycine linker to provide structural flexibility. The TIE^{XVIII} domain, previously described for making multispecific antibodies (Alvarez-Cienfuegos et al., 2016), allows the production of efficiently secreted soluble trimeric proteins by inter- and intramolecular trimerization. Since the combination of 4-1BB and OX40 stimulation has shown synergistic effects in preclinical models (Lee et al., 2004; Qui et al., 2011), a hexavalent 4-1BB $\times$ OX40 bispecific molecule was developed by fusing a 4-1BB and an OX40 targeting Anticalin protein to the TIE^{XVIII} domain (unpublished data). The potential of CAR T cells armored with Anticalin proteins is currently investigated in relevant preclinical models.

Single-chain variable fragments (scFv) are most commonly used as antigen binding moieties in the extracellular CAR domain. However, the stability and/or folding properties of scFvs can render CARs susceptible to oligomerization, clustering, and ligand-independent tonic signaling which has been associated with CAR T cell exhaustion or cell death (Ajina and Maher, 2018; Long et al., 2015; Rafiq et al., 2018). Given their robust structure and stability, Anticalin proteins may have the potential to overcome these limitations and should be further explored as antigen binding domains in CAR molecules to further enhance CAR T cell therapies.

8. Inhalable Anticalin proteins for local lung delivery

Mono- and Duocalin proteins delivered by inhalation are a promising approach in the development of novel biotherapeutics for respiratory disease. Such direct delivery to the lung results in high local drug exposure at the site of the lesion and low systemic exposure avoiding systemic side effects (Bruns et al., 2019a, 2019b). The inhaled mode of application, furthermore, is of high convenience for the patients. In contrast, systemically delivered antibodies generally have a low bioavailability in the lung with only 15% of a systemically administered antibody penetrating the lung tissue (Shah and Betts, 2013). Based on their small size and stable structure, Anticalin proteins exhibit very favorable drug-like properties for inhaled delivery via nebulization or dry powder inhalation (Skerra, 2008). Monoclonal antibodies, for example, are not well suited for inhaled delivery because of limited stability which results in aggregation and loss of activity upon aerosolization or spray drying required for dry powder inhalation (Kane et al., 2013; Maillet et al., 2008). A first proof-of-concept for the inhaled delivery of an Anticalin protein and local target engagement in the lung was demonstrated by a Ph1b clinical trial of PRS-060/AZD1402, a Monocalin blocking the IL-4 receptor alpha (IL-4Ra). Inhibition of fractional exhaled nitric oxide (FeNO), a pulmonary biomarker of lung inflammation, was demonstrated with low dose PRS-060/AZD1402 with no systemic PD activity, measured by inhibition of pSTAT6, a marker for systemic target engagement, suggesting that pulmonary target engagement may be sufficient to reduce airway inflammation (Bruns et al., 2019b). PRS-060/AZD1402 is currently undergoing a phase 2a clinical trial for treatment of moderate to severe asthma (NCT04643158).

Approved biotherapeutics for the treatment of chronic respiratory diseases are monospecific and address only a single pathway, although a plethora of disease driving pathways have been identified. For this reason, combinatorial targeting of multiple pathways, for example by using bispecific molecules, is considered a very promising approach for future drug development for respiratory diseases, including, for example, idiopathic pulmonary fibrosis (discussed in Sato et al., 2019). As described earlier, the modularity of the Anticalin protein technology allows for the generation of bispecific or biparatopic molecules by genetic fusion of two Anticalin proteins (Schlehuber and Skerra, 2001; Siegemund et al., 2021). When compared to other bispecific modalities, Duocalins are of a considerably small size (~40 kDa), consist of only one single polypeptide chain, and feature a high stability intrinsic to Anticalin proteins, thus rendering them a very attractive modality for the development of innovative and inhalable bispecific therapies for chronic respiratory diseases.

9. Concluding remarks

The last decades have seen the rise of protein-based therapies, but also revealed challenges and limitations that require novel formats and designs tailored to target biology and disease in their unique context. To this end, Anticalin proteins greatly enrich the therapeutic design space as a unique class of drug candidates offering high flexibility, potency, and precision. As Mono- or Duocalins, their size and stability allow deep tissue penetration as well as transition over barriers as seen in the case of inhalable Anticalin proteins, which show a high degree of on-target organ activity while minimizing risk of systemic toxicity. High potency and precision can also be achieved in systemic applications by flexible fusion to antibodies, antibody fragments or described means of half-life extension. Such fusion constructs allow multispecificity and multivalency with a control over geometry and proximity of binding sites. Anticalin proteins provide excellent properties as building blocks in terms of stability, developability and manufacturing. The degree of flexibility offered by such, small single-chain scaffolds as monomers, multimers or other fusion formats can provide key advantages in future drug concept design by meeting the challenges that reside in the complexity of immunotherapy of cancer and chronic inflammation. With their small coding size and short intrinsic half-life their application spectrum extends to vectorized approaches such as CAR T cell cargo. Such therapeutic delivery routes can be expected to gain importance in the future as the technologies further develop.

References

Ajina, A., Maher, J., 2018. Strategies to address chimeric antigen receptor tonic signaling. Mol. Cancer Ther. 17, 1795–1815.

Alvarez-Cienfuegos, A., et al., 2016. Intramolecular trimerization, a novel strategy for making multispecific antibodies with controlled orientation of the antigen binding domains. Sci. Rep. 6, 28643.

Arnaud, M., Bobisse, S., Chiffelle, J., Harari, A., 2021. The promise of personalized TCR-based cellular immunotherapy for cancer patients. Front. immunol. 12.
Blanco, B., Domínguez-Alonso, C., Alvarez-Vallina, L., 2021. Bispecific immunomodulatory antibodies for Cancer immunotherapy. Clin. Cancer Res. 27, 5457–5464.
Braud, A., Hannauer, M., Mislin, G.L.A., Schalk, I.J., 2009. The Pseudomonas aeruginosa Pyochelin-Iron uptake pathway and its metal specificity. J. Bacteriol. 191, 3517–3525.
Breustedt, D.A., Korndörfer, I.P., Redl, B., Skerra, A., 2005. The 1.8-Å crystal structure of human tear Lipocalin reveals an extended branched cavity with capacity for multiple ligands. J. Biol. Chem. 280, 484–493.
Breustedt, D.A., Chatwell, L., Skerra, A., 2009. A new crystal form of human tear lipocalin reveals high flexibility in the loop region and induced fit in the ligand cavity. Acta Crystallogr. D 65, 1118–1125.
Bruns, I., Fitzgerald, M., Pardali, K., Gardiner, P., Keeling, D., Axelsson, L., Jiang, F., Lickliter, J., Close, D., 2019a. First-in-human data for the inhaled IL-4Ra antagonist, AZD1402/PRS-060, reveals a promising clinical profile for the treatment of asthma. In: D101. Clinical and translational studies in asthma and COPD. American Thoracic Society, p. A7476.
Bruns, I., et al., 2019b. Late breaking abstract - multiple ascending dose study of the inhaled IL-4Ra antagonist, AZD1402/PRS-060, in mild asthmatics demonstrates robust FeNO reduction and a promising clinical profile for the treatment of asthma. In: Airway pharmacology and treatment. European Respiratory Society, p. PA3709.
Claus, C., et al., 2019. Tumor-targeted 4-1BB agonists for combination with T cell bispecific antibodies as off-the-shelf therapy. Sci. Transl. Med. 11, eaav5989.
Compte, M., et al., 2021. An fc-free EGFR-specific 4-1BB-agonistic Trimerbody displays broad antitumor activity in humanized murine Cancer models without toxicity. Clin. Cancer Res. 27, 3167–3177.
Corvey, C., et al., 2016. Patent—Novel Proteins Specific for Pyoverdine and Pyochelin. World Intellectual Property Organization.
Das, S., Johnson, D.B., 2019. Immune-related adverse events and anti-tumor efficacy of immune checkpoint inhibitors. J. Immunother. Cancer 7, 306.
Dauner, M., Skerra, A., 2020. Scavenging bacterial Siderophores with engineered Lipocalin proteins as an alternative antimicrobial strategy. Chembiochem 21, 601–606.
Deuschle, F.-C., Morath, V., Schiefner, A., Brandt, C., Ballke, S., Reder, S., Steiger, K., Schwaiger, M., Weber, W., Skerra, A., 2020. Development of a high affinity Anticalin ® directed against human CD98hc for theranostic applications. Theranostics 10, 2172–2187.
Deuschle, F.-C., Ilyukhina, E., Skerra, A., 2021. Anticalin® proteins: from bench to bedside. Expert Opin. Biol. Ther. 21, 509–518.
Fisher, T.S., et al., 2012. Targeting of 4-1BB by monoclonal antibody PF-05082566 enhances T-cell function and promotes anti-tumor activity. Cancer Immunol. Immunother. 61, 1721–1733.
Flo, T.H., Smith, K.D., Sato, S., Rodriguez, D.J., Holmes, M.A., Strong, R.K., Akira, S., Aderem, A., 2004. Lipocalin 2 mediates an innate immune response to bacterial infection by sequestrating iron. Nature 432, 917–921.
Gille, H., Huelsmeyer, M., Christian, H.-J., Matschiner, G., Trentmann, S., Skerra, A., Hohlbaum, A., 2008. Discovery and characterization of a novel Anticalin with potent in vivo antagonistic activity towards VEGF-A. Cancer Res. 68. 4077 LP – 4077.
Gille, H., Hülsmeyer, M., Trentmann, S., Matschiner, G., Christian, H.J., Meyer, T., Amirkhosravi, A., Audoly, L.P., Hohlbaum, A.M., Skerra, A., 2016. Functional characterization of a VEGF-A-targeting Anticalin, prototype of a novel therapeutic human protein class. Angiogenesis 19, 79–94.

Goetz, D.H., Holmes, M.A., Borregaard, N., Bluhm, M.E., Raymond, K.N., Strong, R.K., 2002. The neutrophil Lipocalin NGAL is a bacteriostatic agent that interferes with Siderophore-mediated Iron acquisition. Mol. Cell 10, 1033–1043.

Hinner, M.J., et al., 2019. Tumor-localized costimulatory T-cell engagement by the 4-1BB/HER2 bispecific antibody-anticalin fusion PRS-343. Clin. Cancer Res. 25, 5878–5889.

Hodi, F.S., et al., 2010. Improved survival with Ipilimumab in patients with metastatic melanoma. N. Engl. J. Med. 363, 711–723.

Hohlbaum, A.M., et al., 2018. Sustained plasma hepcidin suppression and iron elevation by Anticalin-derived hepcidin antagonist in cynomolgus monkey. Br. J. Pharmacol. 175, 1054–1065.

Hurov, K., et al., 2021. BT7480, a novel fully synthetic bicycle tumor-targeted immune cell agonist™ (bicycle TICA™) induces tumor localized CD137 agonism. J. Immunother. Cancer 9, e002883.

Jackson, H.J., Rafiq, S., Brentjens, R.J., 2016. Driving CAR T-cells forward. Nat. Rev. Clin. Oncol. 13, 370–383.

Jain, T., et al., 2017. Biophysical properties of the clinical-stage antibody landscape. Proc. Natl. Acad. Sci. U. S. A. 114, 944–949.

Jan, M., et al., 2021. Reversible ON- and OFF-switch chimeric antigen receptors controlled by lenalidomide. Sci. Transl. Med. 13.

Jenkins, R.W., Barbie, D.A., Flaherty, K.T., 2018. Mechanisms of resistance to immune checkpoint inhibitors. Br. J. Cancer 118, 9–16.

Kane, C., Neil, K.O., Conk, M., Picha, K., 2013. Inhalation delivery of protein therapeutics. Inflamm. Allergy Drug Targets, 81–87.

Lee, S.-J., Myers, L., Muralimohan, G., Dai, J., Qiao, Y., Li, Z., Mittler, R.S., Vella, A.T., 2004. 4-1BB and OX40 dual Costimulation synergistically stimulate primary specific CD8 T cells for robust effector function. J. Immunol. 173, 3002–3012.

Long, A.H., et al., 2015. 4-1BB costimulation ameliorates T cell exhaustion induced by tonic signaling of chimeric antigen receptors. Nat. Med. 21, 581–590.

Ma, J., Mo, Y., Tang, M., Shen, J., Qi, Y., Zhao, W., Huang, Y., Xu, Y., Qian, C., 2021. Bispecific antibodies: From research to clinical application. Front. Immunol. 12.

Maillet, A., et al., 2008. Aerodynamical, immunological and pharmacological properties of the anticancer antibody cetuximab following nebulization. Pharm. Res. 25, 1318–1326.

Mardiana, S., et al., 2017. A multifunctional role for adjuvant anti-4-1BB therapy in augmenting antitumor response by chimeric antigen receptor T cells. Cancer Res. 77, 1296–1309.

Masuda, Y., et al., 2018. Generation and characterization of a novel small biologic alternative to proprotein convertase subtilisin/kexin type 9 (PCSK9) antibodies, DS-9001a, albumin binding domain–fused anticalin protein. J. Pharmacol. Exp. Ther 365, 368–378.

Maude, S.L., et al., 2014. Chimeric antigen receptor T cells for sustained remissions in leukemia. N. Engl. J. Med. 371, 1507–1517.

McNamara, J.O., Kolonias, D., Pastor, F., Mittler, R.S., Chen, L., Giangrande, P.H., Sullenger, B., Gilboa, E., 2008. Multivalent 4-1BB binding aptamers costimulate CD8+ T cells and inhibit tumor growth in mice. J. Clin. Investig. 118, 376–386.

Melero, I., Bach, N., Hellström, K.E., Aruffo, A., Mittler, R.S., Chen, L., 1998. Amplification of tumor immunity by gene transfer of the co-stimulatory 4-1BB ligand: synergy with the CD28 co-stimulatory pathway. Eur. J. Immunol. 28, 1116–1121.

Melero, I., Murillo, O., Dubrot, J., Hervás-Stubbs, S., Perez-Gracia, J.L., 2008. Multi-layered action mechanisms of CD137 (4-1BB)-targeted immunotherapies. Trends Pharmacol. Sci. 29, 383–390.

Mitchell, M.J., Billingsley, M.M., Haley, R.M., Wechsler, M.E., Peppas, N.A., Langer, R., 2021. Engineering precision nanoparticles for drug delivery. Nat. Rev. Drug Discov. 20, 101–124.

Mross, K., Richly, H., Fischer, R., Scharr, D., Büchert, M., Stern, A., Gille, H., Audoly, L.P., Scheulen, M.E., 2013. First-in-human phase I study of PRS-050 (Angiocal), an Anticalin targeting and antagonizing VEGF-A, in patients with advanced solid tumors. PloS One 8, e83232.
Mullard, A., 2021. FDA approves 100th monoclonal antibody product. Nat. Rev. Drug Discov. 20, 491–495.
Oh, D.Y., Bang, Y.J., 2020. HER2-targeted therapies — a role beyond breast cancer. Nat. Rev. Clin. Oncol. 17, 33–48.
Park, J.H., et al., 2018. Long-term follow-up of CD19 CAR therapy in acute lymphoblastic leukemia. N. Engl. J. Med. 378, 449–459.
Peper-Gabriel, J.K., et al., 2022. The PD-L1/4-1BB bispecific antibody-Anticalin fusion protein PRS-344/S095012 elicits strong T-cell stimulation in a tumor-localized manner. Clinical Cancer Research. clincanres.2762.2021.
Perez-Ruiz, E., et al., 2019. Prophylactic TNF blockade uncouples efficacy and toxicity in dual CTLA-4 and PD-1 immunotherapy. Nature.
Petersen, C.T., et al., 2018. Improving T-cell expansion and function for adoptive T-cell therapy using ex vivo treatment with PI3Kδ inhibitors and VIP antagonists. Blood Adv. 2, 210–223.
Pieris Pharmaceuticals Inc, 2015. Pieris pharmaceuticals' collaborator presents promising preclinical data on tetraspecific infectious disease Anticalin program at ASM conference. Press Release. Available at: https://www.pieris.com/investors/news-events/press-releases/detail/511/pieris-pharmaceuticals-collaborator-presents-promising. Accessed December 1, 2021.
Piha-Paul, S., et al., 2021. Abstract CT017: clinical and biomarker activity of PRS-343, a bispecific fusion protein targeting 4-1BB and HER2, from a phase 1 study in patients with advanced solid tumors (study PRS-343-PCS_04_16). In: Clinical Trials. American Association for Cancer Research. CT017–CT017.
Qui, H.Z., et al., 2011. CD134 plus CD137 dual Costimulation induces Eomesodermin in CD4 T cells to program cytotoxic Th1 differentiation. J. Immunol. 187, 3555–3564.
Rafiq, S., et al., 2018. Targeted delivery of a PD-1-blocking scFV by CAR-T cells enhances anti-tumor efficacy in vivo. Nat. Biotechnol. 36, 847–858.
Renders, L., et al., 2019. First-in-human phase I studies of PRS-080#22, a hepcidin antagonist, in healthy volunteers and patients with chronic kidney disease undergoing hemodialysis. PLos one 14, e0212023.
Richter, A., Eggenstein, E., Skerra, A., 2014. Anticalins: exploiting a non-Ig scaffold with hypervariable loops for the engineering of binding proteins. FEBS Lett. 588, 213–218.
Rothe, C., Skerra, A., 2018. Anticalin® proteins as therapeutic agents in human diseases. BioDrugs 32, 233–243.
Sato, S., Yanagihara, T., Kolb, M.R.J., 2019. Therapeutic targets and early stage clinical trials for pulmonary fibrosis. Expert Opin. Investig. Drugs 28, 19–28.
Schiefner, A., Skerra, A., 2015. The menagerie of human Lipocalins: a natural protein scaffold for molecular recognition of physiological compounds. Acc. Chem. Res. 48, 976–985.
Schlehuber, S., Skerra, A., 2001. Duocalins: engineered ligand-binding proteins with dual specificity derived from the Lipocalin fold. Biol. Chem. 382.
Schlehuber, S., Skerra, A., 2005. Lipocalins in drug discovery: from natural ligand-binding proteins to 'anticalins'. Drug Discov. Today 10, 23–33.
Schonfeld, D., et al., 2009. An engineered lipocalin specific for CTLA-4 reveals a combining site with structural and conformational features similar to antibodies. Proc. Natl. Acad. Sci. 106, 8198–8203.
Schrand, B., Berezhnoy, A., Brenneman, R., Williams, A., Levay, A., Kong, L.Y., Rao, G., Zhou, S., Heimberger, A.B., Gilboa, E., 2014. Targeting 4-1BB costimulation to the tumor stroma with bispecific aptamer conjugates enhances the therapeutic index of tumor immunotherapy. Cancer Immunol. Res. 2, 867–877.

Schuster, S.J., et al., 2017. Chimeric antigen receptor T cells in refractory B-cell lymphomas. N. Engl. J. Med. 377, 2545–2554.

Segal, N.H., et al., 2017. Results from an integrated safety analysis of Urelumab, an agonist anti-CD137 monoclonal antibody. Clin. Cancer Res. 23, 1929–1936.

Shah, D.K., Betts, A.M., 2013. Antibody biodistribution coefficients. MAbs 5, 297–305.

Siegemund, M., et al., 2021. Pharmacokinetic engineering of OX40-blocking Anticalin proteins using monomeric plasma half-life extension domains. Front. Pharmacol. 12.

Skerra, A., 2000. Lipocalins as a scaffold. Biochim. Biophys. Acta -Protein Structure and Molecular Enzymol. 1482, 337–350.

Skerra, A., 2008. Alternative binding proteins: Anticalins - harnessing the structural plasticity of the lipocalin ligand pocket to engineer novel binding activities. FEBS J 275, 2677–2683.

Waldman, A.D., Fritz, J.M., Lenardo, M.J., 2020. A guide to cancer immunotherapy: from T cell basic science to clinical practice. Nat. Rev. Immunol. 20, 651–668.

Yeku, O.O., Purdon, T.J., Koneru, M., Spriggs, D., Brentjens, R.J., 2017. Armored CAR T cells enhance antitumor efficacy and overcome the tumor microenvironment. Sci. Rep. 7.

Yi, M., Niu, M., Xu, L., Luo, S., Wu, K., 2021. Regulation of PD-L1 expression in the tumor microenvironment. J. Hematol. Oncol. 14, 10.

Zhao, D., Klempner, S.J., Chao, J., 2019. Progress and challenges in HER2-positive gastroesophageal adenocarcinoma. J. Hematol. Oncol 12, 1–13.

Zou, W., Wolchok, J.D., Chen, L., 2016. PD-L1 (B7-H1) and PD-1 pathway blockade for cancer therapy: mechanisms, response biomarkers, and combinations. Sci. Transl. Med. 8.

CHAPTER FIVE

Overcoming the limitations of cytokines to improve cancer therapy

Claudia Augusta Di Trani[a,b], Assunta Cirella[a,b], Leire Arrizabalaga[a,b], Myriam Fernandez-Sendin[a,b], Angela Bella[a,b], Fernando Aranda[a,b,†], Ignacio Melero[a,b,c,d,†], and Pedro Berraondo[a,b,c,*,†]

[a]Program of Immunology and Immunotherapy, Cima Universidad de Navarra, Pamplona, Spain
[b]Navarra Institute for Health Research (IDISNA), Pamplona, Spain
[c]Centro de Investigación Biomédica en Red de Cáncer (CIBERONC), Madrid, Spain
[d]Department of Oncology, Clínica Universidad de Navarra, Pamplona, Spain
*Corresponding author: e-mail address: pberraondol@unav.es

Contents

Abstract

Cytokines are pleiotropic soluble proteins used by immune cells to orchestrate a coordinated response against pathogens and malignancies. In cancer immunotherapy, cytokine-based drugs can be developed potentiating pro-inflammatory cytokines or blocking immunosuppressive cytokines. However, the complexity of the mechanisms of action of cytokines requires the use of biotechnological strategies to minimize systemic toxicity, while potentiating the antitumor response. Sequence mutagenesis, fusion proteins and gene therapy strategies are employed to enhance the half-life in circulation, target the desired bioactivity to the tumor microenvironment, and to optimize the therapeutic window of cytokines. In this review, we provide an overview of the different strategies currently being pursued in pre-clinical and clinical studies to make the most of cytokines for cancer immunotherapy.

† Co-senior authors.

International Review of Cell and Molecular Biology, Volume 369
ISSN 1937-6448
https://doi.org/10.1016/bs.ircmb.2022.05.002

1. Introduction

The immune system is composed of cells and soluble factors that create a network responsible for restoring homeostasis in the face of a pathogenic threat. Cytokines are the most relevant soluble factors that mediate communication between the different cell types of the immune system, regulating all the aspects of the innate and adaptive immune responses (Berraondo et al., 2019b; Lin and Karin, 2007). Being soluble proteins, they can be easily administered exogenously or blocked by monoclonal antibodies. Therefore, strategies to potentiate cytokines that promote antitumor responses or to block immunosuppressive cytokines are the most relevant strategies in cancer immunotherapy (Berraondo et al., 2019b; Opal and DePalo, 2000; Perez-Ruiz et al., 2019; Rodriguez-Ruiz et al., 2019). After years of intense pre-clinical and clinical research, only two cytokines have been approved by the Food and Drug Administration (FDA) for the treatment of different malignancies: interferon-alpha (IFNα) represents the first agent for cancer immunotherapy ever approved by the FDA in 1986 for hairy cell leukemia (Golomb et al., 1986), and later for chronic myelogenous leukemia (CML) (Italian Cooperative Study Group on Chronic Myeloid L et al., 1994), follicular non-Hodgkin lymphoma (Solal-Celigny et al., 1993), melanoma (Kirkwood et al., 1996) and AIDS-related Kaposi's sarcoma (Groopman et al., 1984). Interleukin 2 (IL-2) received approval in 1992 for the treatment of renal cell carcinoma (Fyfe et al., 1995) and in 1998 for metastatic melanoma (Atkins et al., 1999).

The approval of these cytokines represented a milestone in cancer immunotherapy (Gutterman, 1994; Negrier et al., 1998). However, the fact that more than 30 years later no new cytokine-based drugs have been approved for clinical use illustrates the difficulties in harnessing the safety and full therapeutic potential of these soluble mediators. To increase the number of cytokine-based drugs, new biotechnology strategies are being implemented based on an increasing understanding of the biology of each cytokine as a result of the study of their synthesis, regulation, signaling and structure.

Cytokines are defined as polypeptides responsible for the function, growth and differentiation of different cell types. All cytokines are polypeptides or glycoproteins of relatively low molecular size (<30 kDa). The IL-12 family is the exception to the rule comprising only heterodimeric cytokines, with a molecular size of up to 70 kDa, in the case of IL-12p70 itself.

The transcription and production of cytokines are usually induced by a stimulus and their expression, in most cases, is also regulated at a post-transcriptional level of their messenger RNAs (mRNAs) (Anderson, 2009; Caput et al., 1986; Maeda and Akira, 2017). The low half-life in circulation and transient expression limits systemic activity, and hence cytokines mainly act in an autocrine or paracrine way. Another strategy to limit the radius of action is that certain cytokines can act anchored to the membrane, as in the case of transforming growth factor beta (TGFβ) (Yang et al., 2013) or IL-15. In this case, anchoring occurs through binding to one of the receptor chains, a phenomenon known as *transpresentation* (Sandau et al., 2004). However, some cytokines can have an endocrine effect through circulation in vesicles such as exosomes (Braicu et al., 2015).

Cytokines exert their action through interaction with high-affinity surface receptors. There are three main signaling pathways used by cytokines. Type I and type II cytokine receptors transduce the signal through the JAK-STAT pathway. Receptor-associated JAK proteins approach the cytoplasmatic tail of the receptor upon cytokine binding, initiating a phosphorylation-mediated process that eventually leads to phosphorylation and dimerization of STAT proteins (O'Shea et al., 2015; Shuai and Liu, 2003). The STAT homodimers or heterodimers are able to translocate to the nucleus and interact with specific binding sites on the promoters or enhancers of target genes. A second important signaling route is the PI3K-AKT pathway (Manning and Cantley, 2007). Cytokine binding activates the intracellular signal transducer PI3K, which converts phosphatidylinositol (3,4)-bisphosphate (PIP2) lipids to phosphatidylinositol (3,4,5)-trisphosphate (PIP3) at the plasma membrane. Binding of AKT to PIP3 triggers a process that leads to the protein kinase AKT activation. The third cytokine signaling route is the MAPK/ERK pathway (Kyriakis and Avruch, 2001). The binding of the cytokine to the receptor allows the Ras protein to exchange GDP for GTP. This active form of Ras triggers a cascade of phosphorylations of MAPK kinases leading to the nuclear translocation and functional control of transcription factors.

The result of the activation of these pathways is the modulation of the expression of hundreds of genes that will be overexpressed or repressed. This gene modification results in the modulation of proliferation rates, differentiation and/or precise functions.

Depending on the specific characteristic of each cytokine, the strategies to develop cytokine-based drugs in cancer therapeutics rely on the following three axes: (*i*) modulation of the activity; (*ii*) increase of the half-life in

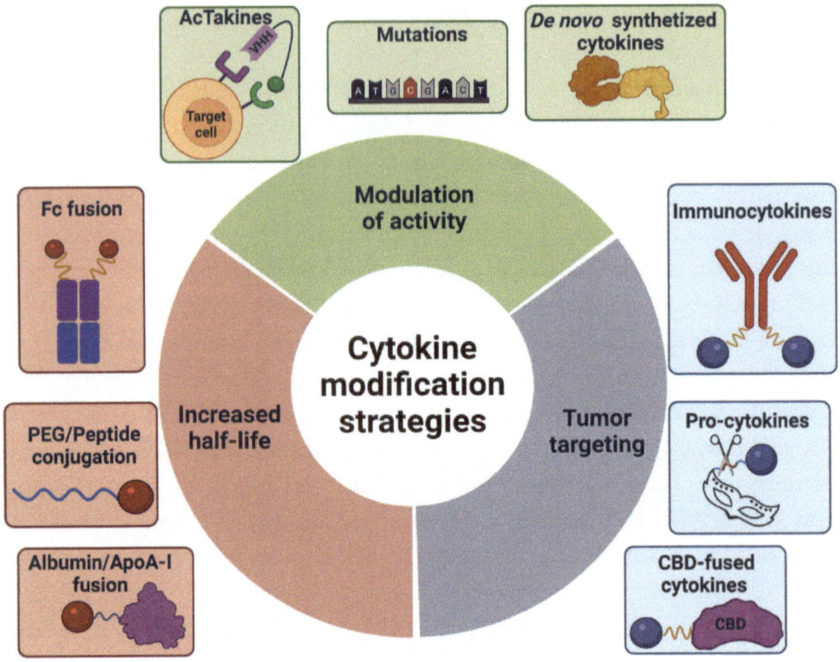

Fig. 1 Strategies to optimize cytokines-based therapies. To develop cytokine-based drugs, modifications are introduced in the cytokine structure in order to modulate the interaction with the cytokine receptors, to target the cytokine to the tumor microenvironment and/or to increase the half-life in circulation. AcTakines = Activity-on-Target cytokines; CBD = collagen-binding domain; Apo-A-I = apolipoprotein A-I; PEG = polyethylene glycol; Fc = fragment crystallizable.

circulation and *iii*) targeting to the tumor microenvironment (TME) (Fig. 1). Novel cytokine-based drugs entering clinical trials harbor at least one modification that leads to the optimization of one of these axes (Propper and Balkwill, 2022). Alternatively, optimization in these axes can be achieved by expressing the cytokine *in vivo* with gene therapy vectors, lipid nanoparticles (LNPs) containing modified mRNA or *ex vivo* modified T cells.

2. Strategies to modulate the activity

Several strategies have been developed to increase and/or tune the activity of cytokines. This can be achieved by the introduction of mutations in their sequences or by fusion to fragments of their receptors or antibodies with the aim of (*i*) increasing or decreasing the affinity to their receptors to modulate agonistic activity, (*ii*) preventing the binding to decoy receptors and (*iii*) re-directing their activity to specific cell types.

One of the cytokines that has been most extensively optimized for cancer immunotherapy is IL-2. This cytokine is a key lymphocyte growth factor, being able to expand and boost the cytolytic activity of T lymphocytes and natural killer (NK) cells (Spolski et al., 2018). The strategies to optimize IL-2 rely on a thorough understanding of the structure and functions of its receptors. The intermediate-affinity dimeric receptor consists of an IL-2Rβ (or CD122) and a common γ chain (γc or CD132), whereas the high-affinity trimeric receptor also includes IL-2Rα (CD25). Dimeric receptors are mainly expressed on cytotoxic T cells and NK cells, while trimeric receptors are predominantly found on regulatory T cells (Tregs). The high-affinity of the trimeric receptor results in preferential stimulation of Tregs that constitutively express CD25, leading to immunosuppression in the tumor microenvironment (Liao et al., 2013). This effect constitutes a detrimental phenomenon for cancer treatment. Moreover, IL-2 broad pleiotropic actions limit its effectiveness due to systemic toxicity that includes vascular leak syndrome, characterized by lung edema and liver damage (Chen et al., 2018). Thus, harnessing the desirable properties of IL-2 while minimizing its counterproductive effects is an important immunotherapeutic goal.

Site-directed mutagenesis approaches have been mostly used to decrease CD25 binding affinity, generating the so-called IL-2 "muteins". These muteins have an antitumor efficacy comparable to wild-type (wt) IL-2 but possess greater therapeutic index in murine models and in non-human primates (Carmenate et al., 2013; Chen et al., 2018; Vazquez-Lombardi et al., 2017).

Yeast and phage displays were successfully used to generate IL-2 variants with increased affinity to one or more receptor subunits. The Wittrup and Garcia groups pioneered the development of such IL-2 variants, which promote the preferential expansion of cytotoxic T cells and NK cells (Levin et al., 2012; Liu et al., 2009). Levin A.M. et al. generated a "superkine" using *in vitro* evolution and achieved an increased binding affinity for IL-2Rβ, and a receptor-binding structure mimicking that acquired when IL-2 is already bound to CD25, but irrespective of CD25 expression, thereby resulting in a more potent phosphorylation of STAT5. The triggered signaling robustly activated cytotoxic T cells and NK cells with reduced toxicity compared to wt IL-2. All these mechanisms led to improved antitumor responses *in vivo*. This mutant, called H9, was further modified by the Warren J.L. group to harness only the desired properties of the cytokine. They introduced a single mutation at the H9–IL-2Rγ interface to promote the expansion of $CD8^+$ T cells without driving terminal differentiation (Mo et al., 2021). Such feature is considered of interest for adoptive T cell therapy (Kishton et al., 2021).

Silva et al. used computational methods to generate a *de novo* synthesized cytokine, a hyper stable and non-immunogenic protein called neoleukin-2/15 (neo-2/15). It is able to bind human and mouse IL-2Rβ/γc with higher affinity than the natural cytokines triggering the downstream signaling in an IL-2Rα- and IL-15Rα- independent manner. The synthetic chimeric cytokine showed greater antitumor activity compared to wt IL-2 in melanoma and colon carcinoma murine models, with reduced toxicity as compared to wt IL-2 (Silva et al., 2019).

IL-15, like IL-2, is a cytokine belonging to the four-α-helix bundle family of cytokines, and has been extensively modified to improve its antitumor activity. IL-2 and IL-15 share two receptor subunits (β and γ chains), whereas the α chain is responsible for the high affinity and confers specificity to one or the other. This cytokine is also a potent growth factor for T lymphocytes and NK cells but with reduced activity on Tregs. The most common mechanism by which IL-15 exerts such effects is called trans-presentation: on the membrane of antigen-presenting cells, IL-15 is tethered to the α chain (IL-15Rα), which offers the cytokine to neighboring T and NK cells which express the γ and β chains, upon cell-to-cell contact (Dubois et al., 2002). Thanks to the deep understanding of its biology, IL-15 has been engineered to introduce mutations to increase the affinity toward γ and β chains (Zhu et al., 2009), to fuse it to Fc domains, to apolipoprotein A-I, to tumor-targeting antibodies and to the sushi domain. The sushi motif is part of the extracellular domain of IL-15Rα that, when fused to IL-15, mimics the transpresentation process required for the full activity of this cytokine giving rise to a superagonist (Mortier et al., 2006; Olsen et al., 2007; Stoklasek et al., 2006; Tan and Waldmann, 2020).

Another cytokine that has been carefully optimized is IL-18. This is a member of the IL-1 cytokine family able to stimulate innate and antigen-experienced lymphocytes but not naïve T cells (Guo et al., 2012). IL-18 activity is regulated by its decoy receptor IL-18 binding protein (IL-18BP). This protein is a secreted antagonist that binds IL-18 with extremely high affinity ($K_D < 1$ nM) (Dinarello et al., 2013). IL-18BP has been found to be elevated 10- to 100-fold in the serum of patients treated with recombinant IL-18 (Robertson et al., 2008), thus neutralizing the potent activity of IL-18 and reducing its efficacy in clinical trials.

In order to shield its antitumoral potential, Zhou et al. used directed evolution with yeast surface display to discover and recombinantly express a "decoy-resistant" IL-18 variant (DR-18), which does not bind to IL-18BP, thus retaining its activity in circulation resulting in better antitumor activity in melanoma and colorectal murine models as compared to wild-type IL-18. Interestingly, DR-18 administration in tumor-bearing mice resulted in a selective expansion of stem-like $TCF1^{+}CD8^{+}$ T cells in the tumor and redirected differentiation toward polyfunctional effector cells rather than toward an exhausted phenotype. Moreover, DR-18 is able to overcome the loss of major histocompatibility complex (MHC) class I by boosting NK cell function, thus constituting a good candidate for treatments of tumors that develop resistance to immunotherapy (Zhou et al., 2020).

AcTakines (Activity-on-Target cytokines) consist of the fusion of the mutant version of a cytokine (with reduced affinity for its receptor) to a nanobody targeting a cell-specific surface marker. Consequently, the fused attenuated cytokines are inactive while on their way to the target but become functional when reaching it. The goal is to eliminate the unfavorable side and sink effects (Tzeng et al., 2015), but having the full activity of the cytokine restored on the targeted cells as a result of local avidity-driven receptor binding (Garcin et al., 2014). This clever strategy has been applied in preclinical models to IFNα directed to tumor necrosis factor receptor 1 (TNFR1)-expressing cells and conventional dendritic cells (cDC1) (Cauwels et al., 2018; Garcin et al., 2014; Tomasello et al., 2014) and to TNF and IFNγ directed to the tumor neovasculature (Huyghe et al., 2020; Uze and Tavernier, 2015).

3. Strategies to increase the half-life

Cytokine half-life in circulation is usually short, restricting their radius of action to the production site. In part, short half-lives are explained by being proteins under the glomerular filtration threshold (~60 kDa). For this reason, cytokine-based therapies need multiple infusions to achieve the desired therapeutic concentration in the target tissue, the other side of the coin is that this is responsible for cytokine-induced systemic toxicity. One strategy to improve the pharmacokinetic profile consists in the fusion of the therapeutic cytokine to a molecule that extends the half-life in circulation. In this context, the most relevant system in the clinic consists of the attachment of linear or branched polyethylene glycol (PEG) to the cytokine. PEGylation represents a milestone in pharmaceutical research that has been used for commercial protein drugs since the 1990s (Ekladious et al., 2019; Nischan and Hackenberger, 2014). PEGylation is routinely achieved by the incubation of a reactive derivative of PEG with the target molecule. The attachment of PEG to the therapeutic molecule shields it from the host's immune system, reduces proteolysis, and increases its solubility and molecular weight, thus reducing plasma clearance by renal filtration.

Site-specific PEGylation overcomes the problems associated with the heterogeneity of the products from the first-generation approaches, achieving the precise conjugation to the desired sites. All these mechanisms result in improved pharmacokinetics that compensates for the reduction of activity of the cytokine due to the binding of PEG molecules to residues critical for activity.

To date, only one PEGylated IFNα2b (PEG-IFN) has been approved for cancer, as adjuvant treatment of melanoma patients in 2011 (Herndon et al., 2012), but many pre-clinical and clinical studies have been performed with other PEGylated cytokines and a few examples are listed below.

Bempegaldesleukin (BEMPEG or NKTR-214) is a prodrug consisting of multiple releasable PEG chains attached to IL-2. On the one hand, the chemical modification permits the sustained and controlled delivery of IL-2, increasing its serum half-life. On the other hand, the site-specific attachment of PEG at the interface between the cytokine and CD25 dampens the interaction with such subunit, without interfering with the interaction with the intermediate affinity receptor. This strategy results in the specific stimulation of $CD8^+$ T cells and NK cells without expanding intratumoral Tregs (Charych et al., 2016, 2017). Interestingly, the combination of this prodrug with nivolumab demonstrated antitumor activity in patients with different kinds of solid tumors (NCT02983045, NCT03785925, NCT04410445, NCT03635983, NCT03729245, NCT04540705, NCT04209114). Bempegaldesleukin is also being evaluated in combination with pembrolizumab in solid metastatic tumors (NCT04969861, NCT03138889) and, additionally, as a potential treatment for mild COVID-19 (NCT04646044).

Another cytokine that has benefited from PEGylation is IL-10. In pre-clinical models, PEGylated IL-10, pegilodecakin, induced intratumoral $CD8^+$ T cell infiltration, secretion of cytotoxic enzymes, and IFNγ resulting in higher antigen presentation and tumor growth control (Mumm et al., 2011). A phase I study (NCT02009449) demonstrates that PEGylation of IL-10 extends the plasma half-life from <5 to 20 h. Moreover, daily subcutaneous administration of pegilodecakin resulted in stable serum concentrations for the duration of treatment, sustained immune stimulation, and evidence of antitumor activity, particularly in renal cell cancer (Naing et al., 2016). It has also been combined with pembrolizumab or nivolumab, showing a manageable toxicity profile and promising antitumor activity (Naing et al., 2019).

As shown in these few examples, PEGylation has many advantages, including enhanced drug solubilization, controlled delivery, increased efficacy and improved pharmacokinetics. Nonetheless, one of the major challenges is finding the balance between maximal shielding of the protein while keeping its full bioactivity. Nonspecific conjugation as well as the nondegradable nature of PEG have generated increasing concerns (Cong et al., 2012). To address these issues, biodegradable and non-immunogenic alternatives to PEG have emerged, such as amino acid-based synthetic polypeptides. Several platforms have been developed based on different polypeptides such as XTEN, ELP, PAS, and PEP (Strohl, 2015). These platforms have demonstrated that the conjugation of synthetic polypeptides to therapeutic cytokines represents a viable alternative to PEGylation, opening up new opportunities for the development of responsive release systems for cytokine therapeutics.

The intricate biology of the neonatal fragment crystallizable receptor (FcRn) represents a unique opportunity for developing systems to extend half-life of cytokines in circulation. FcRn is ubiquitously expressed and has a high affinity to the Fc domain of the immunoglobulins G (IgGs) (Ghetie and Ward, 2000). The interaction of these molecules with the FcRn rescues them from intracellular degradation, recycling the internalized molecule (Martin et al., 2001; Ober et al., 2004). Moreover, it increases the molecular mass reducing renal clearance. Therefore, the fusion of the cytokine of interest with the Fc domain extends their half-life in circulation (Ghetie et al., 1997).

Importantly, the Fc domain interacts both with Fc γ receptors (FcγRs) expressed on leucocytes and with the C1 complement protein, inducing the antibody-dependent cell cytotoxicity and phagocytosis (ADCC and ADCP) and complement-dependent cytotoxicity (CDC), respectively (Jefferis and Lund, 2002). These phenomena can be prevented through different approaches explained in the immunocytokines section, further in the text. Alternatively, they can be exploited for a therapeutical advantage, as showed in the following example.

Vazquez-Lombardi et al., in 2017, fused both wild-type IL-2 and a mutated IL-2 (to avoid CD25 binding) to the Fc domain of an IgG2c. As expected, this resulted in increased half-life of both molecules but, surprisingly, wild-type IL-2-Fc resulted in more potent antitumoral efficacy in murine melanoma and colorectal carcinoma models. The authors discovered that such efficacy was due to the depletion of the immunosuppressive Tregs, due to the fact that the wild-type cytokine preferentially bound those cells inducing the ADCC mediated by the fused Fc (Vazquez-Lombardi et al., 2017).

Another cytokine with outstanding antitumoral properties is IL-12, although it is accompanied by almost as many toxicities (Cirella et al., 2022). This heterodimeric cytokine comprises two subunits of 40 and 35 kDa linked by a disulfide bond. In 2018, Jung et al. generated a new construct by fusing the hIgG4-Fc to IL-12, increasing the half-life of the recombinant IL-12 from 5 to 10 h up the 66 h, in the case of their constructs. Particularly, they constructed two IL-12-Fc molecules, one carrying one p40 and one p35 (mono-IL-12) and the other carrying two single-chain IL-12 moieties (bi-IL-12), where the p40 is joined to the p35 through a flexible aminoacidic linker. Surprisingly, mono-mIL-12-Fc proved to be more potent than bi-mIL-12-Fc in rescuing CT26-HER2/neu-bearing mice, through the triggering of IFNγ production, the cytotoxicity of effector cells as well as a more efficient induction of long-lived memory $CD8^{+}$ T cells (Jung et al., 2018).

Albumin is a globular monomeric plasma protein of 67 kDa which is also an ideal candidate to be used as a carrier of therapeutic cytokines. It lacks enzymatic or immunological functions, increases the molecular weight of the cytokine, and has a circulation half-life of 19 days in humans. The latter is due to a recycling process mediated by the FcRn, similar to that observed for IgGs (Andersen et al., 2012; Chaudhury et al., 2003, 2006). The fusion of albumin to cytokines has been applied to enhance the antitumor efficacy of IL-2 and G-CSF, none of which have been granted clinical approval as yet (Gladkov et al., 2016; Skrombolas and Frelinger, 2014).

Another approach used to improve the pharmacokinetics of cytokines is the fusion to apolipoprotein A-I (ApoA-I). ApoA-I is the main protein component of the high-density lipoproteins that transport cholesterol from tissues to the liver (Francis, 2010). Apolipoprotein A-I fused to cytokines increases the half-life in circulation, and the biodistribution is determined by the scavenger receptor, class B type 1 (SR-B1). As a consequence of this interaction, the fused cytokine accumulates not only in the liver but also in tumors, making these fusion proteins especially suitable for cancer immunotherapy (Fioravanti et al., 2011, 2012; Medina-Echeverz et al., 2014; Ochoa et al., 2013, 2018; Vasquez et al., 2016, 2017b).

Hence, half-life of cytokines can be increased through chemical or genetic engineering, and each of the different approaches has its own advantages and drawbacks.

4. Strategies to target the tumor microenvironment

The systemic administration of pro-inflammatory cytokines can lead to severe adverse events which hinder their clinical application. One approach to reducing systemic toxicity is to fuse the therapeutic payload to a tumor-targeting moiety, such as antibodies, with the aim of localizing the activity of the cytokine where it is most needed. In the case of the immunocytokines, two aspects are of pivotal importance: the choice of the antigen(s) to target and the antibody format.

Finding an antigen that is uniquely expressed by tumor cells or by the tumor microenvironment is the Holy Grail of cancer treatment. The compromise which researchers usually accept is to choose an antigen that is overexpressed in the tumor niche with respect to healthy tissues. To date, none of the immunocytokines has obtained FDA approval, but many are being evaluated in clinical trials (Table 1), as a result of hundreds of pre-clinical studies performed over the past decades (Murer and Neri, 2019). The targets of the immunocytokines thus far tested in patients include molecules expressed in the tumor vasculature, tumor-specific gangliosides, oncofetal antigens, necrotic tissue, tumor-associated fibroblasts, immune checkpoints or tumor-overexpressed proteins.

Regarding the antibody format, an important issue concerns the Fc domain as this is responsible for off-target effects due to the interaction with the FcγRs and C1q, as previously described. In order to minimize such effects, different strategies have been developed such as choosing the antibody isotype with a lower affinity to its receptors (human IgG4 and murine IgG1) or deglycosylating the Fc fragment by treatment with endoglycosidases. Alternatively, the glycosylation can be prevented by rendering Fc fragments "silent" by inserting point mutations in the constant heavy chain domain 2 (CH2) sequence of the antibodies to reduce Fc affinity to its receptors (reviewed in Liu et al., 2020). These approaches abrogate binding to FcγRs and C1q while maintaining the interaction with the FcRn responsible for the long-life in circulation.

The other option consists in completely eliminating the Fc domain from the therapeutical construct. In fact, intense efforts have been made to design Fc-devoid antibody fragments that retain their fragment antigen-binding

Table 1 Immunocytokines in clinical trials.

Name	Ab format	Cytokine	Target	Indication	Clinical trial ID and status
DI-Leu16-IL-2	L Kappa; IgG1	wt IL-2	CD20	B-cell NHL	Phase I completed: NCT00720135 Phase I/II terminated: NCT02151903 NCT01874288
Hu14.18-IL-2	L Kappa; IgG1	wt IL-2	GD2	Melanoma	Phase I completed: NCT00003750 Phase I/II suspended: NCT03958383 Phase II completed: NCT00590824 NCT00109863
				Neuroblastoma and osteosarcoma	Phase I suspended: NCT03209869
				Neuroblastoma	Phase II completed: NCT01334515 NCT00082758
RO6874281 (RG7461; Simlukafusp alfa)	Human IgG1	IL-2v: F42A, Y45A, L72G to abolish CD25 binding	FAP	Breast cancer and head and neck cancer	Phase I active n.r.: NCT02627274
				Renal cell carcinoma	Phase I completed: NCT03063762
				Head and neck, oesophageal and cervical cancers	Phase II, completed: NCT03386721
				Melanoma	Phase I active n.r.: NCT03875079
RO6895882 (Cergutuzumab amunaleukin)	Humanized IgG1	IL-2v: F42A, Y45A, L72G to abolish CD25 binding	CEA	Solid CEA + tumors	Phase I completed: NCT02004106 NCT02350673

EMD52183 (Selectikine)	Human IgG2	IL-2v: D20T selectivity for activated T cells and reduced binding to endothelial cells.	DNA/histone complex	NSCLC	Phase I completed: NCT00879866
				Solid tumors and B-cell NHL	Phase I completed: NCT01032681
EMD27306 (Tucotuzumab celmoleukin)	Humanized IgG	wt IL-2	EpCAM	Ovarian and prostate cancers, CRC or NSCLC	Phase I completed: NCT00132522
				Kidney, bladder, or lung cancer	Phase I completed: NCT00016237
RO7284755	Human IgG	IL-2v: to to abolish CD25 binding	PD-1	Solid tumors	Phase I recruiting: NCT04303858
NHS-IL-12 (M7824)	Human IgG1	IL-12	DNA/histone complex	Breast cancer	Phase I recruiting: NCT04756505
				Solid tumors	Phase I completed: NCT01417546
				Prostate tumor	Phase I/II recruiting: NCT04633252
				Genitourinary malignancies	Phase I recruiting: NCT04235777
				Advanced Kaposi sarcoma	Phase I/II recruiting: NCT04303117
BC1-IL-12 (AS1409)	IgG	IL-12	EDB	RCC, melanoma	Phase I completed: NCT00625768

Continued

Table 1 Immunocytokines in clinical trials.—cont'd

Name	Ab format	Cytokine	Target	Indication	Clinical trial ID and status
L19-IL-2 (Darleukin)	Human Diabody	IL-2	EDB	Melanoma	Phase I/II active n.r.: NCT02076646
					Phase II completed: NCT01253096
					Phase II terminated: NCT01055522
				Oligometastatic solid tumor	Phase III recruiting: NCT02938299
					Phase I completed: NCT02086721
				B-cell lymphoma	Phase I/II recruiting: NCT02957019
L19-TNF (Onfekafusp alfa; Fibromun)	Human Diabody	TNF	EDB	Melanoma	Phase II completed: NCT02076633
					Phase III recruiting: NCT02938299
				Colorectal cancer	Phase I/II completed: NCT01253837
				Soft tissue sarcoma	Phase II recruiting: NCT03420014
F16-IL-2 (Teleukin)	Human Diabody	IL-2	Tnc A1	AML	Phase I active n.r.: NCT02957032
					Phase I completed: NCT03207191
				Solid tumors	Phase I/II terminated: NCT01131364
					Phase I/II completed: NCT01134250
				Merkel cell carcinoma	Phase II terminated: NCT02054884

The table summarizes all the immunocytokines that are being tested in clinical trials. The first column presents the name or names of the molecule, the second column contains the published information about the antibody format. Moreover, we indicate the cytokine employed and if it presents any mutations and the target of the antibody. The last two columns indicate in which malignancies the molecule is being tested with the relative unique identification code with the status of the trial, if known. *Abbreviations.* L: light chain; n.r.: not recruiting; wt: wild-type; IL-2v: Interleukin-2 Variant; B-cell NHL: B-cell Non-Hodgkin Lymphoma; NSCLC: Non Small-Cell Lung Cancer; CRC: Colorectal cancer; RCC: Renal cell carcinoma; AML: acute myeloid leukemia; GD2: disialoganglioside; FAP: fibroblast activation protein; CEA: carcinoembryonic antigen. PD-1: Programmed Death-ligand 1; EDB: alternatively-spliced extradomain B of fibronectin; Tnc A1: extra-domain A1 of tenascin-C.

(Fab) arms and thus their targeting specificity. The most common include classic monovalent antibody fragments (Fab and single-chain fragment variable (scFv)), engineered variants (diabodies, triabodies, tetrabodies, minibodies) and single-domain antibodies (VHH or nanobodies). These Fc-devoid formats offer many advantages over the full-size 150 kDa heavy IgGs, such as increased tumor penetration (Sanz et al., 2005), the capacity to bind to epitopes inaccessible for the large and flat antigen-binding sites of conventional antibodies (Klein et al., 2009; Rossey et al., 2017), abrogation of the Fc-dependent off-target effects and easier large-scale production. But every advantage comes with a drawback that, in this case, concerns the pharmaceutical performance of the molecules which have a much shorter half-life in circulation. These molecules are not recycled by the FcRn, and, if smaller than 60 kDa, are filtrated up by the kidney and removed from circulation. Thus, despite the great promise of small antibodies formats, most of the therapeutic approaches currently used in clinical trials use full-size antibodies, and those which were first designed as Fc-devoid have now included Fc domains in order to extend their half-life (reviewed in Sanz and Alvarez-Vallina, 2021).

Another approach used to reduce the on-target off-tumor effects is to generate pro-cytokines that can be selectively activated in the TME. The strategies implemented so far consist of fusing the cytokines open reading frame (ORF) to the sequence of the latency-associated peptide (LAP) from TGFβ (Mullen et al., 2014), to an scFv specific for the cytokine, to a masking polypeptide or to its native receptors. In order to make the cytokine active only at the site of disease, researchers exploit the overexpression of matrix metalloproteinases (MMPs) in inflamed tissues and in the TME by engineering a specific cleavage site for MMPs between the masking component and the active cytokine. Mullen et al. used the LAP strategy with different cytokines and tested in experimental autoimmune encephalomyelitis murine models (Mullen et al., 2014).

The scFv-based strategy was developed by Skrombolas et al. These authors generated a fusion protein consisting of an IL-12-specific scFv linked to the heterodimeric protein through an MMP9-cleavable peptide. *In vitro,* the full activity of the IL-12 is restored only when the peptide is proteolyzed by MMPs, and therefore the cytokine is free to bind its receptor and trigger the downstream signaling (Skrombolas et al., 2019).

Wang et al. designed a synthetic polypeptide-IFNα conjugate linked through an MMP2-responsive sequence (Wang et al., 2019b). The resulting conjugate, defined PEP20-M-IFN, can therefore release the therapeutical cytokine achieving both protein activation and deep penetration into the TME, compared to the MMP sequence-devoid conjugate. These researchers demonstrated that systemic administration of PEP20-M-IFN exerted a potent antitumor efficacy against both OVCAR-3 and SKOV-3 ovarian tumor models in mice. The same group engineered a body temperature-responsive and MMP-cleavable conjugate of IFNα and elastin-like polypeptide that enabled a spatiotemporally programmed two-step release kinetics. The authors were able to improve the antitumor efficacy compared with the native IFNα in two mouse models of melanoma and ovarian cancer (Wang et al., 2019a).

With the objective of localizing the potent IL-12 to the TME, in January 2022, Xue et al. improved the IL-12-Fc molecule by fusing to the p40 moiety the extracellular domain of the IL-12Rβ1, known as the chain primarily responsible for high-affinity binding to the cytokine, thus exploiting its native high specificity and eliminating the risk of introducing any potential immunogenic epitopes (Xue et al., 2022). The aminoacidic linker sequence contains a motif specific for the cleavage by MMP14, known to be upregulated in many tumoral tissues as compared to adjacent healthy tissues (Turunen et al., 2017). This pro-IL-12, when injected systemically into tumor-bearing mice, leads to lower systemic toxicity compared to IL-12-Fc while still retaining, or in some tumor models even outperforming, the antitumoral efficacy in murine tumor models in a $CD8^+$ and IFNγ-dependent manner. After demonstrating the capacity of the molecule to overcome tyrosine kinase inhibitor- and immune checkpoint blockade (ICB)-related resistances in murine tumoral models, they generated the human counterpart of the molecule thus establishing a strong base for its potential future development in clinical testing (Xue et al., 2022).

As already mentioned, finding the best target(s) is the focus of continuous efforts aimed at enhancing the localization of the therapeutic cytokine in the tumor microenvironment. One promising target is collagen, a major component of the extracellular matrix (Ricard-Blum, 2011). It has been observed that many tumors contain increased amounts of collagen as compared to healthy tissues (Butcher et al., 2009; Zhou et al., 2017). Moreover, the pathological structure of tumor vessels makes them more permeable than those from healthy tissues (Maeda et al., 2000). Thus, cytokines fused to collagen-binding domains (CBDs) exploit the increased expression of collagens and tumor-specific accessibility. Researchers from the Hubbell group fused IL-2 to a CBD peptide with a high affinity to collagen I and III (Ishihara et al., 2019). The fusion protein increased the tumor infiltration of $CD8^+$ T cells and induced a superior antitumor efficacy as compared to the unmodified counterparts.

CBD-conjugation was further used for the tumor-targeted delivery of several cytokines, leading to more efficient and safer cancer immunotherapies. In a recent report by the Wittrup group (Momin et al., 2019), IL-12 and IL-2 were fused to lumican, a protein that binds collagen I and IV. Intratumoral injection of the fusion proteins resulted in cytokines anchoring to the tumor extracellular matrix and achieved not only local curative efficacy but also abscopal effects on non-injected tumor lesions. Similarly, Mansurov et al. evaluated the systemic administration of CBD-IL-12, which resulted in a higher accumulation of the cytokine in murine breast tumors and lower systemic levels compared to the unmodified counterpart. The treatment provided greater anticancer efficacy, eliciting complete regression of checkpoint inhibitor-unresponsive breast tumors (Mansurov et al., 2020).

The high potential of this promising strategy is limited by the quantity of collagen available in the TME, which is characteristic of each patient and the type of tumor. To improve such methodology, Agarwal Y. et al. produced a single-chain IL-12 bound to aluminum hydroxide (alum), which is an FDA-approved vaccine adjuvant (Agarwal et al., 2022). The strategy exploits the fact that alum aggregates form a physical depot at injection sites in tissue that is persistent for several weeks (Flarend et al., 1997). Indeed, a single intratumoral dose of alum-anchored IL-12 tethered the cytokine to the tumor for 6 days, eliciting a safer and stronger IFNγ-dependent systemic antitumor response in poorly immunogenic murine tumor models, compared to unanchored IL-12.

5. *In vivo* production of cytokines

Viral and non-viral methods to transfer genetic material encoding for cytokines constitute an alternative to attain the desired activity of cytokines in cancer therapeutics (Fig. 2).

The rationale of the strategy consists of using optimized cytokines and/or exploiting the biology of the chosen vector, both of which can provide therapeutic advantages. In fact, depending on the vector characteristics, long-term expression or local expression in the tumor microenvironment can be achieved. Long-lasting expression can be achieved using lentivirus, third-generation adenovirus or adeno-associated vectors (Berraondo et al., 2005; Crettaz et al., 2009; Gonzalez-Aparicio et al., 2011). However, the expression of cytokines must be regulated to avoid toxicity due to the long-term expression of the cytokine (Vanrell et al., 2011;

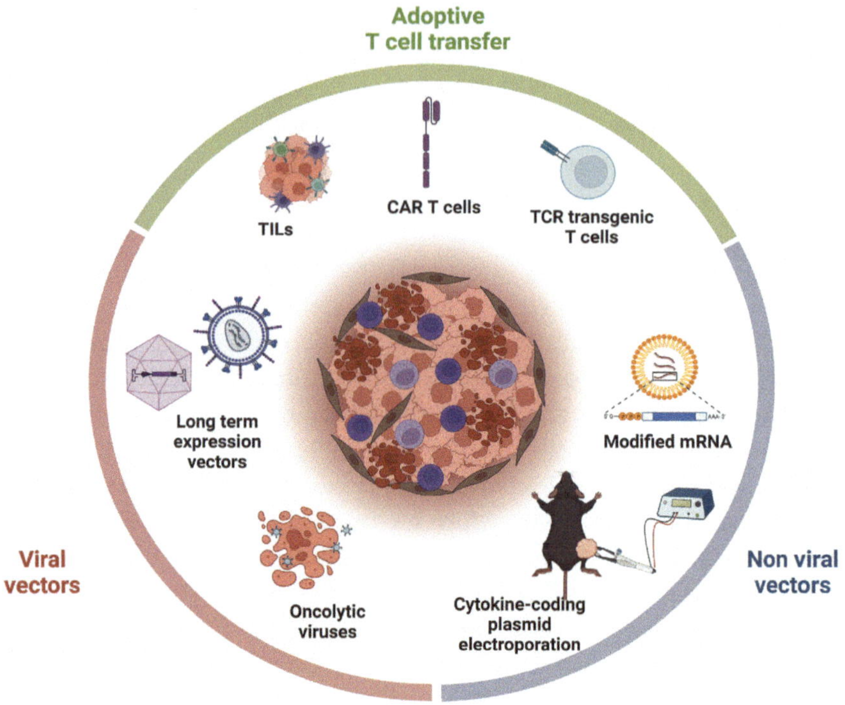

Fig. 2 *In vivo* production of cytokines. The production of the therapeutic cytokine can be achieved *in vivo* by adoptive transfer of cytokine-secreting engineered T cells, by cytokine-coding viral vectors or non-viral methods such as mRNAs and *via* electroporation of plasmidic DNA. TILs = tumor infiltrating lymphocytes; CAR = chimeric antigen receptor; TCR = T cell receptor; mRNAs = messenger RNAs.

Vasquez et al., 2017a). Chiocca et al. generated a replication-incompetent adeno-associated virus (AAV) containing human IL-12p70 sequence downstream of an oral drug-inducible promoter (Chiocca et al., 2019). The authors conducted a phase I trial in patients undergoing resection of recurrent high-grade glioma (NCT02026271). The intratumoral injection of this AAV resulted in a drug dose-dependent increase in IL-12 and IFNγ levels. Side effects, such as cytokine release syndrome, correlated to the drug dose and were reversible upon drug administration discontinuation. The approach is extremely appealing, but further optimization is required. The complexity of implementing regulatable systems in the clinic has favored the vectors that allow short-term expression or vectors that promote an *in situ* vaccine effect due to the tumor cell killing while releasing the cytokine of interest (Sanchez-Paulete et al., 2018).

One method that has recently gained momentum is delivery in oncolytic viruses. An oncolytic virus is a virus that specifically infects and lyses cancer cells but not normal cells. This ability can occur naturally or can be provided in the laboratory by engineering natural viruses. These modifications started a new era of less toxic cancer-targeted virotherapy (Hemminki and Hemminki, 2016). Such modifications are intended to: enhance virus tropism to cancer cells, restrict replication to cancer cells or enhance efficacy by inserting therapeutic transgenes. Indeed, they can be armed with pro-inflammatory cytokines (Russell et al., 2012). Talimogene laherparepvec (T-VEC) is the only oncolytic virus approved by the FDA for the treatment of cancer. It is a genetically modified non-virulent herpes simplex virus that encodes GM-CSF. The intratumoral administration in melanoma lesions demonstrated overall response benefit due to effects in treated lesions but also in distant non-treated lesions (Andtbacka et al., 2015).

Other oncolytic viruses armed with cytokines are being evaluated in clinical trials (Macedo et al., 2020), but the interest in non-viral vectors is increasing exponentially. These vectors have the advantage of reduced immunogenicity as compared to viral vectors and circumvent the risk of insertional mutagenesis. Indeed, cytokine-coding plasmid electroporation and modified mRNA delivered naked or complexed in lipid nanoparticles have reached clinical testing.

The first attempts to electroporate cytokine-coding plasmids date back to the beginning of this century and were conducted with plasmids encoding for IL-2 and GM-CSF or with IL-2 and IL-12 in murine melanoma models (Heller et al., 2000; Lohr et al., 2001). The extremely promising pre-clinical results with IL-12 led to clinical testing in 2008, under the name of tavokinogene telseplasmid (Berraondo et al., 2018). Intratumoral IL-12-coding plasmid electroporation therapy proved to be efficacious in a phase I study of metastatic melanoma (NCT00323206), leading to the evaluation in phase II clinical trials in patients with different tumors as a single treatment or combined with checkpoint inhibitors (NCT01502293, NCT02345330, NCT01440816, NCT03823131, NCT04526730, NCT03567720, NCT03132675) (Daud et al., 2008; Greaney et al., 2020). Results from these recently completed clinical studies have demonstrated a local immune response and, subsequently, a systemic effect as either a monotherapy or combination treatment approach.

In the case of mRNA-based therapeutics, the therapeutic payload consists of an *in vitro* transcribed mRNA that is delivered *in vivo* by lipid nanoparticles (LNP), electroporation or a specific saline buffer. On entry into the cytoplasm of the target cell, the therapeutic cytokine is translated, post-translationally modified, and secreted (if required) by the host cell machineries. This strategy bypasses all the issues related to recombinant protein production, such as purification, stability, and post-translational modifications (Berraondo et al., 2019a; Di Trani et al., 2022; Jiang et al., 2018; Pastor et al., 2018).

The other important advantage of mRNA-based therapeutics is that the intrinsic transitory nature of mRNA makes the potential side effects of the cytokine self-limiting. The short kinetics, together with the feasibility of repeated administration of the mRNAs, makes the repeated expression of the cytokine or the mixture of synergistic cytokines feasible.

The undiscussed potential and applicability of such therapeutics became clear to the whole world in December 2020, with the first approval ever of mRNA-based nanomedicines with COVID-19 vaccines. This achievement represents only the tiny tip of the iceberg, what lies underwater is the incessant research performed over the last three decades focusing on improving, on the one hand, the mRNA sequence and, on the other, its delivery. mRNA sequence optimization focused especially on the study of the untranslated regulatory regions and on the need to insert (or not) modified nucleoside in the mRNA sequence (Holtkamp et al., 2006; Kariko et al., 2005; Sahin et al., 2014). The delivery efficacy of mRNA can be improved by rational design of lipids through modulation of head groups and hydrophobic tails to increase cellular uptake and endosomal escape of lipid nanoparticle–mRNA formulations (Hou et al., 2021) or it can also be delivered naked in saline solutions. As a result, three mRNA-based therapies are being evaluated for the treatment of solid tumors.

MEDI1191 is a lipid nanoparticle-formulated mRNA encoding a monomeric bioactive IL-12p70 protein which gave impressive results in murine tumor models (Hewitt et al., 2020) and is currently being evaluated in a phase I clinical trial as a monotherapy and in combination with durvalumab in patients with advanced solid tumors. The LNP-mRNA is administered by intratumoral injection, while the anti-PD-L1 antibody is injected intravenously (NCT03946800). The ongoing testing shows that the local treatment is well-tolerated and offers encouraging advantages in terms of antitumor efficacy (Hamid et al., 2021).

The possibility of simultaneous expression of synergistic cytokines is being exploited in two clinical trials. mRNA-2752 is an LNP-formulated mRNA encoding human OX40L, IL-23 and IL-36γ cytokines. This combination proved to be synergistic in pre-clinical mouse tumor models, in a cross-presenting DCs- and cytotoxic $CD8^+$ T cells-dependent manner

(Hewitt et al., 2019). A phase I clinical trial is being performed to evaluate the intratumoral administration of mRNA-2752 alone or combined with systemic durvalumab in patients with advanced malignancies (NCT03739931). The treatments have been shown to be tolerable at all dose levels studied and to be associated with tumor shrinkage (Bauer et al., 2019; Patel et al., 2020).

SAR441000 is a saline-formulated mRNA mixture encoding human single-chain IL-12, IFNα, GM-CSF, and IL-15 fused to the sushi domain of IL-15Rα. Following promising pre-clinical results (Hotz et al., 2021), a first-in-human, dose-escalation and expansion study is being performed to evaluate the intratumoral administration of the mRNA cocktail as monotherapy or in combination with cemiplimab in patients with advanced solid tumors (NCT03871348) (Bechter et al., 2020).

Another interesting application of *in vivo* expressed cytokines consists of the *ex vivo* translation/transduction of immune cells that are subsequently adoptively transferred to cancer patients. The expression of cytokines can improve any of the three modalities of adoptive T cell therapy: tumor-infiltrating T lymphocytes (TILs), T cell receptor (TCR) transgenic T lymphocytes or chimeric antigen receptor (CAR) T cells. Again, many efforts have focused on IL-2 because the *ex vivo* expanded T cells require the concomitant administration of this cytokine, usually administered intravenously at the highest tolerated dose every 8 h to support the expansion and function of the adoptively transferred T or NK cells (Rosenberg et al., 2008). In order to redirect IL-2 activities only to transferred T cells and to elude undesired activities on the other cell types, Sockolosky et al. engineered IL-2 cytokine-receptor "orthogonal pairs" that interact with each other but do not interact with their natural cytokine and receptor counterparts (Sockolosky et al., 2018). Introduction of ortho-IL-2Rβ into T cells enabled the selective targeting of ortho-IL-2 to engineered $CD4^+$ and $CD8^+$ T cells *in vitro* and *in vivo* resulting in an antitumor response in B16-F10 melanoma tumor models, with limited off-target effects and toxicity. These orthogonal pairs have also been introduced into CD19-CAR T cells augmenting the antitumor activity of the adoptive transfer in leukemia murine and xenograft models (Aspuria et al., 2021; Zhang et al., 2021).

In 2021, Agliardi et al. treated orthotopic syngeneic murine model of glioma with systemic infusion of EGFRvIII-CAR T cells accompanied by a single intratumoral injection of IL-12-Fc. The combination with the cytokine reshaped the myeloid and T cell compartments leading to the eradication of advanced tumors, whereas either single treatment alone failed to control tumor growth (Agliardi et al., 2021).

Fourth-generation CARs, also called TRUCKs (T-cells redirected for universal cytokine-mediated killing) (Chmielewski et al., 2014), compared to the previous CAR generations, are additionally engineered to secrete a transgenic cytokine upon CAR signaling in the targeted tumor tissue. Technically, the induced protein expression is under control of the NFAT6-IL-2 minimal promoter which is activated upon CAR signaling, and this makes the cytokine secretion to specifically occur in the TME (Chmielewski and Abken, 2020).

To date, the release of transgenic cytokines by CAR T cells has been reported for IL-18 and was found to augment the effect of CD19-CAR T cells in a xenograft model (Hu et al., 2017). IL-18 was combined with carcinoembryonic antigen (CEA)-CAR T cells against murine pancreatic tumors, inducing a T-Bet^{high} FoxO1^{low} effector signature in the CAR T cells and promoting the regression of advanced tumors (Chmielewski and Abken, 2017). It has also been combined with gp100-specific TCRα/β against murine melanoma tumors (Kunert et al., 2017).

IL-12 has also been tested to armor CEA-CAR T cells. The cytokine release in the TME prevented the antigen-loss of cancer cells and the immune evasion, induced the infiltration of activated macrophages resulting in antitumoral efficacy in the targeted CEA^{+} tumors (Chmielewski et al., 2011). Moreover, this potent cytokine was used to empower CD19-CAR T cells either in preconditioned models (Pegram et al., 2012) or in a model with no need for preconditioning (Kueberuwa et al., 2018) to armor GLYPICAN-CAR T cells against murine and human hepatocellular carcinoma (HCC) tumors models (Liu et al., 2019).

Other cytokines have been explored to improve adoptive T cell transfer. For instance, *ex vivo* stimulation of OT-I T cells with IL-15 prior to infusion has been found to induce a memory phenotype of transferred T cells, with respect to the effector phenotype induced by IL-2 stimulation (Xu et al., 2016). Strikingly, CD19-CAR T cells have been engineered with a tethered variant of IL-15 whose STAT5-dependent signaling increased T-cell survival and persistence *in vitro* and *in vivo*. Those long-lived T cells possessed a memory stem-cell transcriptional profile resulting in greater antitumor efficacy (Hurton et al., 2016). Chen et al. engineered T cells with a construct encompassing the sequence of IL-15 and the one for the disialoganglioside (GD2)-CAR separated by a self-cleaving peptide. The produced cytokine during the *ex vivo* expansion culture led to the enrichment of cells with memory and stem cell-like phenotypes, which promoted tumor eradication of neuroblastoma xenograft models (Chen et al., 2019). Recently,

Liu et al. employed IL-15 to engineer cord blood-derived NK cells modified to express a CAR against CD19. IL-15 endowed transferred NK cells with enhanced *in vivo* proliferation, persistence and antitumor activity in a humanized model of leukemia, compared to the cells lacking the cytokine (Liu et al., 2018).

Koneru et al. demonstrated how IL-12 constitutive secreting human MUC16 CAR T cells were more persistent *in vivo* and showed antitumor efficacy and higher systemic IFNγ levels in an ovarian xenograft model (Koneru et al., 2015b). The same group used a syngeneic model of murine ovarian peritoneal carcinomatosis and showed that IL-12 armored MUC16-CAR T cells overcame the inhibitory ascitic microenvironment and PD-L1-mediated inhibition (Yeku et al., 2017). On the basis of these promising results, CAR T cells targeting MUC16 and secreting IL-12 have been tested in a phase I clinical trial (NCT02498912) (Koneru et al., 2015a). A NFAT promoter-driven IL-12 has been used to armor an EGFR-specific CAR T cells for the treatment of metastatic colon cancer in a phase I clinical trial (NCT03542799).

Apart from CAR T cells, IL-12 has been used to armor autologous TILs for the treatment of metastatic melanoma (NCT01236573), but the phase I/II was prematurely terminated due to clinical toxicities. The fact that IL-12 expression was under the NFAT inducible promoter was not enough to limit IL-12-dependent toxicities, such as liver dysfunction, high fevers and hemodynamic instability (Zhang et al., 2015).

In order to add further control on IL-12 expression and bioactivity, Etxeberria et al., in 2019, electroporated tumor-specific T lymphocytes with an IL-12-coding mRNA (Etxeberria et al., 2019). Compared to mock-electroporated cells, the direct injection of OT-I IL-12 cells into murine melanomas induced a systemic immune response able to control both treated and distant non-injected tumor lesions. The transient nature of the electroporated payload achieved such efficacy without the serious toxicities associated with sustained production of this cytokine (Etxeberria et al., 2019; Minute et al., 2020). The efficacy was further increased by combining the therapy with agonists of 4-1BB signaling.

6. Conclusions

We have sought in this review to recapitulate the main strategies employed by researchers and biotech companies to ameliorate cytokine-based therapeutics, that represent one of the many weapons against malignancies. Cancer is a leading cause of death worldwide, and a dysfunctional immune system is intimately associated with tumor development, progression, and recurrence (Vinay et al., 2015). Indeed, tumor escape from immune control is recognized as one of the "hallmarks of cancer" (Hanahan, 2022).

For this reason, immunotherapy has gained increasing attention as a powerful tool for awakening the immune system and reinvigorating it against malignancies. The first immunotherapeutics that were granted with marketing approval for the treatment of various types of tumors have been cytokines, with recombinant IFNα being the first to be used against hairy-cell leukemia almost 40 years ago. Renewed interest in cancer immunotherapy has arisen following the large clinical impact of immune checkpoint blocking antibodies and adoptive transfer of CAR T cells.

Cytokines integrate the innate and adaptive immune responses, representing how immune cells communicate with one another and with cells outside the immune system. Both pro-inflammatory and anti-inflammatory cytokines, by acting on their multiple targets and exerting their many functions, tip the balance in response to an alteration of homeostasis in order to restore it. It is undoubted that these soluble, relatively small proteins represent an important therapeutic arsenal for developing new compounds for the treatment of cancer. The continuous growth of biotechnology has led to the development of numerous clever approaches to overcome the intrinsic limitations of cytokine-based therapies. Such strategies are aimed at increasing cytokine half-life, tethering the cytokine to the tumor microenvironment and improving the pharmacodynamic profiles (as summarized in Fig. 1). One way to reduce cytokine-related systemic toxicity is by the generation of immunocytokines intended at concentrating the cytokine either to the TME or their bioactivity to specific cell types, as in the case of AcTakines. Moreover, cytokines can be fused to the collagen-binding domains as an alternative strategy to anchor their bioactivities in the tumor niche. Additionally, systemic toxicity can be avoided by administering cytokines "masked" by different kinds of moieties in order to be conditionally activated by tumor-associated MMPs. Site-directed mutagenesis, yeast and phage display have been employed to give rise to cytokines or *de novo* cytokines with the desired properties while at the same time eliminating or reducing the undesirable effects. The fusion of the cytokine with other compounds also increases the molecular weight of the payload overcoming another drawback of cytokines, their short half-life due to renal clearance. To this end, they can also be fused to albumin, apolipoprotein A-I, synthetic polypeptides or chemically conjugated with PEG, which represented a milestone in the improvement of drug pharmacokinetics and solubilization. Furthermore, cytokines can be fused to Fc domains of the IgGs and researchers can exploit their affinity to FcγRs and complement proteins or can render them silent, depending on the desired biological activity. We listed these strategies as independent from one another, but in many cases more than one optimization is combined in the same chimera. One of the myriads of examples is the RO7284755, an immunocytokine that is currently being evaluated in a phase I clinical trial (NCT04303858). It encompasses an anti-human PD-1 linked to an engineered, variant form of interleukin-2 (IL-2v), two synergistic molecules are here combined to revert T cell exhaustion. Another example is represented by IL-15 superagonist complex ALT-803, which consists of a mutated version of IL-15, with increased affinity to IL-15Rβ, bound to the sushi domain to mimic transpresentation process and further fused to the Fc domain to increase half-life (159).

Recent advances in the field of gene transfer have allowed these molecules to be synthesized *in vivo*, adding another level of control over cytokine activity. This includes viral vectors such as long-expression gene therapy vectors and oncolytic viruses that, apart from delivering the therapeutic cytokine, also lyse cancer cells. Non-viral methods are currently gaining momentum because they elude the risk of the immunogenicity and insertional mutagenesis associated with viral vectors and such approaches include *in vivo* cytokine-coding plasmid electroporation and mRNA-based drugs, as depicted in Fig. 2. These therapies, by intratumoral injection, allow the expression of the cytokine/mixture of cytokines in the TME for a short period of time, thus overcoming many of the limitations of the recombinant protein production process and the toxicity related to the sustained expression of the cytokine. Other potent vectors that permit the expression of the cytokine directly *in vivo* are adoptively transferred lymphocytes after their *ex vivo* empowerment with the cytokine of interest. Hence cytokines also represent an important tool to potentiate and synergize with other immunotherapies already in clinical practice.

All in all, cytokine-based immunotherapy represents an ever-expanding field that includes a high number of approaches that need an equally wide range of biotechnology and bioinformatic techniques, delivery methods, and intense research efforts and rationales directed at attaining the desired response of the therapeutic cytokine. Thanks to the integration of all the fields together, we envision significant advances in cytokines-based therapies in the next future.

Acknowledgments

We are grateful to Dr. Paul Miller for English editing. This work was supported by Instituto de Salud Carlos III (PI20/00002, and PI19/01128) cofinanced by Fondos FEDER "A way to make Europe" and Joint Translational Call for Proposals 2015 (JTC 2015), TRANSCAN456 2 (code: TRS-2016-00000371), Gobierno de Navarra Proyecto LINTERNA Ref.: 0011-1411-2020-000075. This project has received funding from the European Union's Horizon 2020 research and innovation program under the Marie Skłodowska-Curie grant agreement No 765394. F.A. receives a Miguel Servet I (CP19/00114) contract from ISCIII (Instituto de Salud Carlos III) co-financed by FSE (Fondo Social Europeo) "Investing in your future". Á.B. is recipient of PFIS fellowship from ISCIII (FI20/00058), and L.A. is recipients for a fellowship of the Aid Program Assigned to Projects from the University of Navarra. Some figures contain elements from Biorender (https://biorender.com).

References

Agarwal, Y., Milling, L.E., Chang, J.Y.H., Santollani, L., Sheen, A., Lutz, E.A., et al., 2022. Intratumourally injected alum-tethered cytokines elicit potent and safer local and systemic anticancer immunity. Nat. Biomed. Eng. 6 (2), 129–143.

Agliardi, G., Liuzzi, A.R., Hotblack, A., De Feo, D., Nunez, N., Stowe, C.L., et al., 2021. Intratumoral IL-12 delivery empowers CAR-T cell immunotherapy in a pre-clinical model of glioblastoma. Nat. Commun. 12 (1), 444.

Andersen, J.T., Dalhus, B., Cameron, J., Daba, M.B., Plumridge, A., Evans, L., et al., 2012. Structure-based mutagenesis reveals the albumin-binding site of the neonatal Fc receptor. Nat. Commun. 3, 610.

Anderson, P., 2009. Intrinsic mRNA stability helps compose the inflammatory symphony. Nat. Immunol. 10 (3), 233–234.

Andtbacka, R.H., Kaufman, H.L., Collichio, F., Amatruda, T., Senzer, N., Chesney, J., et al., 2015. Talimogene Laherparepvec improves durable response rate in patients with advanced melanoma. J. Clin. Oncol. 33 (25), 2780–2788.

Aspuria, P.J., Vivona, S., Bauer, M., Semana, M., Ratti, N., McCauley, S., et al., 2021. An orthogonal IL-2 and IL-2Rbeta system drives persistence and activation of CAR T cells and clearance of bulky lymphoma. Sci. Transl. Med. 13 (625), eabg7565.

Atkins, M.B., Lotze, M.T., Dutcher, J.P., Fisher, R.I., Weiss, G., Margolin, K., et al., 1999. High-dose recombinant interleukin 2 therapy for patients with metastatic melanoma: analysis of 270 patients treated between 1985 and 1993. J. Clin. Oncol. 17 (7), 2105–2116.

Bauer, T., Patel, M., Jimeno, A., Wang, D., McDermott, J., Zacharek, S., et al., 2019. Abstract CT210: a phase I, open-label, multicenter, dose escalation study of mRNA-2752, a lipid nanoparticle encapsulating mRNAs encoding human OX40L, IL-23, and IL-36γ, for intratumoral injection alone and in combination with immune checkpoint blockade. Cancer Res. 79 (suppl. 13), CT210.

Bechter, O., Utikal, J., Baurain, J.-F., Massard, C., Sahin, U., Derhovanessian, E., et al., 2020. 391 A first-in-human study of intratumoral SAR441000, an mRNA mixture encoding IL-12sc, interferon alpha2b, GM-CSF and IL-15sushi as monotherapy and in combination with cemiplimab in advanced solid tumors. J. Immunother. Cancer 8 (Suppl 3), A237.

Berraondo, P., Ochoa, L., Crettaz, J., Rotellar, F., Vales, A., Martinez-Anso, E., et al., 2005. IFN-alpha gene therapy for woodchuck hepatitis with adeno-associated virus: differences in duration of gene expression and antiviral activity using intraportal or intramuscular routes. Mol. Ther. 12 (1), 68–76.

Berraondo, P., Etxeberria, I., Ponz-Sarvise, M., Melero, I., 2018. Revisiting Interleukin-12 as a Cancer immunotherapy agent. Clin. Cancer Res. 24 (12), 2716–2718.

Berraondo, P., Martini, P.G.V., Avila, M.A., Fontanellas, A., 2019a. Messenger RNA therapy for rare genetic metabolic diseases. Gut 68 (7), 1323–1330.
Berraondo, P., Sanmamed, M.F., Ochoa, M.C., Etxeberria, I., Aznar, M.A., Perez-Gracia, J.L., et al., 2019b. Cytokines in clinical cancer immunotherapy. Br. J. Cancer 120 (1), 6–15.
Braicu, C., Tomuleasa, C., Monroig, P., Cucuianu, A., Berindan-Neagoe, I., Calin, G.A., 2015. Exosomes as divine messengers: are they the Hermes of modern molecular oncology? Cell Death Differ. 22 (1), 34–45.
Butcher, D.T., Alliston, T., Weaver, V.M., 2009. A tense situation: forcing tumour progression. Nat. Rev. Cancer 9 (2), 108–122.
Caput, D., Beutler, B., Hartog, K., Thayer, R., Brown-Shimer, S., Cerami, A., 1986. Identification of a common nucleotide sequence in the 3′-untranslated region of mRNA molecules specifying inflammatory mediators. Proc. Natl. Acad. Sci. U. S. A. 83 (6), 1670–1674.
Carmenate, T., Pacios, A., Enamorado, M., Moreno, E., Garcia-Martinez, K., Fuente, D., et al., 2013. Human IL-2 mutein with higher antitumor efficacy than wild type IL-2. J. Immunol. 190 (12), 6230–6238.
Cauwels, A., Van Lint, S., Paul, F., Garcin, G., De Koker, S., Van Parys, A., et al., 2018. Delivering type I interferon to dendritic cells empowers tumor eradication and immune combination treatments. Cancer Res. 78 (2), 463–474.
Charych, D.H., Hoch, U., Langowski, J.L., Lee, S.R., Addepalli, M.K., Kirk, P.B., et al., 2016. NKTR-214, an engineered cytokine with biased IL2 receptor binding, increased tumor exposure, and marked efficacy in mouse tumor models. Clin. Cancer Res. 22 (3), 680–690.
Charych, D., Khalili, S., Dixit, V., Kirk, P., Chang, T., Langowski, J., et al., 2017. Modeling the receptor pharmacology, pharmacokinetics, and pharmacodynamics of NKTR-214, a kinetically-controlled interleukin-2 (IL2) receptor agonist for cancer immunotherapy. PLoS One 12 (7), e0179431.
Chaudhury, C., Mehnaz, S., Robinson, J.M., Hayton, W.L., Pearl, D.K., Roopenian, D.C., et al., 2003. The major histocompatibility complex-related Fc receptor for IgG (FcRn) binds albumin and prolongs its lifespan. J. Exp. Med. 197 (3), 315–322.
Chaudhury, C., Brooks, C.L., Carter, D.C., Robinson, J.M., Anderson, C.L., 2006. Albumin binding to FcRn: distinct from the FcRn-IgG interaction. Biochemistry 45 (15), 4983–4990.
Chen, X., Ai, X., Wu, C., Wang, H., Zeng, G., Yang, P., et al., 2018. A novel human IL-2 mutein with minimal systemic toxicity exerts greater antitumor efficacy than wild-type IL-2. Cell Death Dis. 9 (10), 989.
Chen, Y., Sun, C., Landoni, E., Metelitsa, L., Dotti, G., Savoldo, B., 2019. Eradication of neuroblastoma by T cells redirected with an optimized GD2-specific chimeric antigen receptor and interleukin-15. Clin. Cancer Res. 25 (9), 2915–2924.
Chiocca, E.A., Yu, J.S., Lukas, R.V., Solomon, I.H., Ligon, K.L., Nakashima, H., et al., 2019. Regulatable interleukin-12 gene therapy in patients with recurrent high-grade glioma: results of a phase 1 trial. Sci. Transl. Med. 11 (505).
Chmielewski, M., Abken, H., 2017. CAR T cells releasing IL-18 convert to T-bet(high) FoxO1(low) effectors that exhibit augmented activity against advanced solid tumors. Cell Rep. 21 (11), 3205–3219.
Chmielewski, M., Abken, H., 2020. TRUCKS, the fourth-generation CAR T cells: current developments and clinical translation. Expert Opin. Biol. Ther. 3 (3), e84.
Chmielewski, M., Kopecky, C., Hombach, A.A., Abken, H., 2011. IL-12 release by engineered T cells expressing chimeric antigen receptors can effectively muster an antigen-independent macrophage response on tumor cells that have shut down tumor antigen expression. Cancer Res. 71 (17), 5697–5706.

Chmielewski, M., Hombach, A.A., Abken, H., 2014. Of CARs and TRUCKs: chimeric antigen receptor (CAR) T cells engineered with an inducible cytokine to modulate the tumor stroma. Immunol. Rev. 257 (1), 83–90.
Cirella, A., Luri-Rey, C., Di Trani, C.A., Teijeira, A., Olivera, I., Bolaños, E., et al., 2022. Novel strategies exploiting interleukin-12 in cancer immunotherapy. Pharmacol. Ther. 239, 108189.
Cong, Y., Pawlisz, E., Bryant, P., Balan, S., Laurine, E., Tommasi, R., et al., 2012. Site-specific PEGylation at histidine tags. Bioconjug. Chem. 23 (2), 248–263.
Crettaz, J., Otano, I., Ochoa-Callejero, L., Benito, A., Paneda, A., Aurrekoetxea, I., et al., 2009. Treatment of chronic viral hepatitis in woodchucks by prolonged intrahepatic expression of interleukin-12. J. Virol. 83 (6), 2663–2674.
Daud, A.I., DeConti, R.C., Andrews, S., Urbas, P., Riker, A.I., Sondak, V.K., et al., 2008. Phase I trial of interleukin-12 plasmid electroporation in patients with metastatic melanoma. J. Clin. Oncol. 26 (36), 5896–5903.
Di Trani, C.A., Fernandez-Sendin, M., Cirella, A., Segues, A., Olivera, I., Bolanos, E., et al., 2022. Advances in mRNA-based drug discovery in cancer immunotherapy. Expert Opin. Drug Discovery 17 (1), 41–53.
Dinarello, C.A., Novick, D., Kim, S., Kaplanski, G., 2013. Interleukin-18 and IL-18 binding protein. Front. Immunol. 4, 289.
Dubois, S., Mariner, J., Waldmann, T.A., Tagaya, Y., 2002. IL-15Ralpha recycles and presents IL-15 in trans to neighboring cells. Immunity 17 (5), 537–547.
Ekladious, I., Colson, Y.L., Grinstaff, M.W., 2019. Polymer-drug conjugate therapeutics: advances, insights and prospects. Nat. Rev. Drug Discov. 18 (4), 273–294.
Etxeberria, I., Bolanos, E., Quetglas, J.I., Gros, A., Villanueva, A., Palomero, J., et al., 2019. Intratumor adoptive transfer of IL-12 mRNA transiently engineered antitumor CD8(+) T cells. Cancer Cell 36 (6), 613–29 e7.
Fioravanti, J., Gonzalez, I., Medina-Echeverz, J., Larrea, E., Ardaiz, N., Gonzalez-Aseguinolaza, G., et al., 2011. Anchoring interferon alpha to apolipoprotein A-I reduces hematological toxicity while enhancing immunostimulatory properties. Hepatology 53 (6), 1864–1873.
Fioravanti, J., Medina-Echeverz, J., Ardaiz, N., Gomar, C., Parra-Guillen, Z.P., Prieto, J., et al., 2012. The fusion protein of IFN-alpha and apolipoprotein A-I crosses the blood-brain barrier by a saturable transport mechanism. J. Immunol. 188 (8), 3988–3992.
Flarend, R.E., Hem, S.L., White, J.L., Elmore, D., Suckow, M.A., Rudy, A.C., et al., 1997. In vivo absorption of aluminium-containing vaccine adjuvants using 26Al. Vaccine 15 (12 – 13), 1314–1318.
Francis, G.A., 2010. The complexity of HDL. Biochim. Biophys. Acta 1801 (12), 1286–1293.
Fyfe, G., Fisher, R.I., Rosenberg, S.A., Sznol, M., Parkinson, D.R., Louie, A.C., 1995. Results of treatment of 255 patients with metastatic renal cell carcinoma who received high-dose recombinant interleukin-2 therapy. J. Clin. Oncol. 13 (3), 688–696.
Garcin, G., Paul, F., Staufenbiel, M., Bordat, Y., Van der Heyden, J., Wilmes, S., et al., 2014. High efficiency cell-specific targeting of cytokine activity. Nat. Commun. 5, 3016.
Ghetie, V., Ward, E.S., 2000. Multiple roles for the major histocompatibility complex class I- related receptor FcRn. Annu. Rev. Immunol. 18, 739–766.
Ghetie, V., Popov, S., Borvak, J., Radu, C., Matesoi, D., Medesan, C., et al., 1997. Increasing the serum persistence of an IgG fragment by random mutagenesis. Nat. Biotechnol. 15 (7), 637–640.
Gladkov, O., Moiseyenko, V., Bondarenko, I.N., Shparyk, Y., Barash, S., Adar, L., et al., 2016. A phase III study of balugrastim versus pegfilgrastim in breast cancer patients receiving chemotherapy with doxorubicin and docetaxel. Oncologist 21 (1), 7–15.

Golomb, H.M., Jacobs, A., Fefer, A., Ozer, H., Thompson, J., Portlock, C., et al., 1986. Alpha-2 interferon therapy of hairy-cell leukemia: a multicenter study of 64 patients. J. Clin. Oncol. 4 (6), 900–905.

Gonzalez-Aparicio, M., Alzuguren, P., Mauleon, I., Medina-Echeverz, J., Hervas-Stubbs, S., Mancheno, U., et al., 2011. Oxaliplatin in combination with liver-specific expression of interleukin 12 reduces the immunosuppressive microenvironment of tumours and eradicates metastatic colorectal cancer in mice. Gut 60 (3), 341–349.

Greaney, S.K., Algazi, A.P., Tsai, K.K., Takamura, K.T., Chen, L., Twitty, C.G., et al., 2020. Intratumoral plasmid IL12 electroporation therapy in patients with advanced melanoma induces systemic and Intratumoral T-cell responses. Cancer Immunol. Res. 8 (2), 246–254.

Groopman, J.E., Gottlieb, M.S., Goodman, J., Mitsuyasu, R.T., Conant, M.A., Prince, H., et al., 1984. Recombinant alpha-2 interferon therapy for Kaposi's sarcoma associated with the acquired immunodeficiency syndrome. Ann. Intern. Med. 100 (5), 671–676.

Guo, L., Junttila, I.S., Paul, W.E., 2012. Cytokine-induced cytokine production by conventional and innate lymphoid cells. Trends Immunol. 33 (12), 598–606.

Gutterman, J.U., 1994. Cytokine therapeutics: lessons from interferon alpha. Proc. Natl. Acad. Sci. U. S. A. 91 (4), 1198–1205.

Hamid, O., Hellman, M., Carneiro, B., Marron, T., Subbiah, V., Mehmi, I., et al., 2021. 19O preliminary safety, antitumor activity and pharmacodynamics results of HIT-IT MEDI1191 (mRNA IL-12) in patients with advanced solid tumours and superficial lesions. Ann. Oncol. 32, S9.

Hanahan, D., 2022. Hallmarks of cancer: new dimensions. Cancer Discov. 12 (1), 31–46.

Heller, L., Pottinger, C., Jaroszeski, M.J., Gilbert, R., Heller, R., 2000. In vivo electroporation of plasmids encoding GM-CSF or interleukin-2 into existing B16 melanomas combined with electrochemotherapy induces long-term antitumour immunity. Melanoma Res. 10 (6), 577–583.

Hemminki, O., Hemminki, A., 2016. A century of oncolysis evolves into oncolytic immunotherapy. Onco. Targets. Ther. 5 (2), e1074377.

Herndon, T.M., Demko, S.G., Jiang, X., He, K., Gootenberg, J.E., Cohen, M.H., et al., U.S., 2012. Food and Drug Administration approval: peginterferon-alfa-2b for the adjuvant treatment of patients with melanoma. Oncologist 17 (10), 1323–1328.

Hewitt, S.L., Bai, A., Bailey, D., Ichikawa, K., Zielinski, J., Karp, R., et al., 2019. Durable anticancer immunity from intratumoral administration of IL-23, IL-36gamma, and OX40L mRNAs. Sci. Transl. Med. 11 (477).

Hewitt, S.L., Bailey, D., Zielinski, J., Apte, A., Musenge, F., Karp, R., et al., 2020. Intratumoral IL12 mRNA therapy promotes TH1 transformation of the tumor microenvironment. Clin. Cancer Res. 26 (23), 6284–6298.

Holtkamp, S., Kreiter, S., Selmi, A., Simon, P., Koslowski, M., Huber, C., et al., 2006. Modification of antigen-encoding RNA increases stability, translational efficacy, and T-cell stimulatory capacity of dendritic cells. Blood 108 (13), 4009–4017.

Hotz, C., Wagenaar, T.R., Gieseke, F., Bangari, D.S., Callahan, M., Cao, H., et al., 2021. Local delivery of mRNA-encoded cytokines promotes antitumor immunity and tumor eradication across multiple preclinical tumor models. Sci. Transl. Med. 13 (610), eabc7804.

Hou, X., Zaks, T., Langer, R., Dong, Y., 2021. Lipid nanoparticles for mRNA delivery. Nat. Rev. Mater., 1–17.

Hu, B., Ren, J., Luo, Y., Keith, B., Young, R.M., Scholler, J., et al., 2017. Augmentation of antitumor immunity by human and mouse CAR T cells secreting IL-18. Cell Rep. 20 (13), 3025–3033.

Hurton, L.V., Singh, H., Najjar, A.M., Switzer, K.C., Mi, T., Maiti, S., et al., 2016. Tethered IL-15 augments antitumor activity and promotes a stem-cell memory subset in tumor-specific T cells. Proc. Natl. Acad. Sci. U. S. A. 113 (48), E7788–E7797.

Huyghe, L., Van Parys, A., Cauwels, A., Van Lint, S., De Munter, S., Bultinck, J., et al., 2020. Safe eradication of large established tumors using neovasculature-targeted tumor necrosis factor-based therapies. EMBO Mol. Med. 12 (2), e11223.

Ishihara, J., Ishihara, A., Sasaki, K., Lee, S.S., Williford, J.M., Yasui, M., et al., 2019. Targeted antibody and cytokine cancer immunotherapies through collagen affinity. Sci. Transl. Med. 11 (487).

Italian Cooperative Study Group on Chronic Myeloid L, Tura, S., Baccarani, M., Zuffa, E., Russo, D., Fanin, R., et al., 1994. Interferon alfa-2a as compared with conventional chemotherapy for the treatment of chronic myeloid leukemia. N. Engl. J. Med. 330 (12), 820–825.

Jefferis, R., Lund, J., 2002. Interaction sites on human IgG-fc for FcgammaR: current models. Immunol. Lett. 82 (1–2), 57–65.

Jiang, L., Berraondo, P., Jerico, D., Guey, L.T., Sampedro, A., Frassetto, A., et al., 2018. Systemic messenger RNA as an etiological treatment for acute intermittent porphyria. Nat. Med. 24 (12), 1899–1909.

Jung, K., Ha, J.H., Kim, J.E., Kim, J.A., Kim, Y.J., Kim, C.H., et al., 2018. Heterodimeric fc-fused IL12 shows potent antitumor activity by generating memory CD8(+) T cells. Onco. Targets. Ther. 7 (7), e1438800.

Kariko, K., Buckstein, M., Ni, H., Weissman, D., 2005. Suppression of RNA recognition by toll-like receptors: the impact of nucleoside modification and the evolutionary origin of RNA. Immunity 23 (2), 165–175.

Kirkwood, J.M., Strawderman, M.H., Ernstoff, M.S., Smith, T.J., Borden, E.C., Blum, R.H., 1996. Interferon alfa-2b adjuvant therapy of high-risk resected cutaneous melanoma: the Eastern Cooperative Oncology Group Trial EST 1684. J. Clin. Oncol. 14 (1), 7–17.

Kishton, R.J., Vodnala, S.K., Vizcardo, R., Restifo, N.P., 2021. Next generation immunotherapy: enhancing stemness of polyclonal T cells to improve anti-tumor activity. Curr. Opin. Immunol. 74, 39–45.

Klein, J.S., Gnanapragasam, P.N., Galimidi, R.P., Foglesong, C.P., West Jr., A.P., Bjorkman, P.J., 2009. Examination of the contributions of size and avidity to the neutralization mechanisms of the anti-HIV antibodies b12 and 4E10. Proc. Natl. Acad. Sci. U. S. A. 106 (18), 7385–7390.

Koneru, M., O'Cearbhaill, R., Pendharkar, S., Spriggs, D.R., Brentjens, R.J., 2015a. A phase I clinical trial of adoptive T cell therapy using IL-12 secreting MUC-16(ecto) directed chimeric antigen receptors for recurrent ovarian cancer. J. Transl. Med. 13, 102.

Koneru, M., Purdon, T.J., Spriggs, D., Koneru, S., Brentjens, R.J., 2015b. IL-12 secreting tumor-targeted chimeric antigen receptor T cells eradicate ovarian tumors in vivo. Onco. Targets. Ther. 4 (3), e994446.

Kueberuwa, G., Kalaitsidou, M., Cheadle, E., Hawkins, R.E., Gilham, D.E., 2018. CD19 CAR T cells expressing IL-12 eradicate lymphoma in fully lymphoreplete mice through induction of host immunity. Mol. Ther. Oncolytics 8, 41–51.

Kunert, A., Chmielewski, M., Wijers, R., Berrevoets, C., Abken, H., Debets, R., 2017. Intra-tumoral production of IL18, but not IL12, by TCR-engineered T cells is non-toxic and counteracts immune evasion of solid tumors. Onco. Targets. Ther. 7 (1), e1378842.

Kyriakis, J.M., Avruch, J., 2001. Mammalian mitogen-activated protein kinase signal transduction pathways activated by stress and inflammation. Physiol. Rev. 81 (2), 807–869.

Levin, A.M., Bates, D.L., Ring, A.M., Krieg, C., Lin, J.T., Su, L., et al., 2012. Exploiting a natural conformational switch to engineer an interleukin-2 'superkine'. Nature 484 (7395), 529–533.

Liao, W., Lin, J.X., Leonard, W.J., 2013. Interleukin-2 at the crossroads of effector responses, tolerance, and immunotherapy. Immunity 38 (1), 13–25.

Lin, W.W., Karin, M., 2007. A cytokine-mediated link between innate immunity, inflammation, and cancer. J. Clin. Invest. 117 (5), 1175–1183.

Liu, D.V., Maier, L.M., Hafler, D.A., Wittrup, K.D., 2009. Engineered interleukin-2 antagonists for the inhibition of regulatory T cells. J. Immunother. 32 (9), 887–894.

Liu, E., Tong, Y., Dotti, G., Shaim, H., Savoldo, B., Mukherjee, M., et al., 2018. Cord blood NK cells engineered to express IL-15 and a CD19-targeted CAR show long-term persistence and potent antitumor activity. Leukemia 32 (2), 520–531.

Liu, Y., Di, S., Shi, B., Zhang, H., Wang, Y., Wu, X., et al., 2019. Armored inducible expression of IL-12 enhances antitumor activity of glypican-3-targeted chimeric antigen receptor-engineered T cells in hepatocellular carcinoma. J. Immunol. 203 (1), 198–207.

Liu, R., Oldham, R.J., Teal, E., Beers, S.A., Cragg, M.S., 2020. Fc-engineering for modulated effector functions-improving antibodies for cancer treatment. Antibodies (Basel) 9 (4).

Lohr, F., Lo, D.Y., Zaharoff, D.A., Hu, K., Zhang, X., Li, Y., et al., 2001. Effective tumor therapy with plasmid-encoded cytokines combined with in vivo electroporation. Cancer Res. 61 (8), 3281–3284.

Macedo, N., Miller, D.M., Haq, R., Kaufman, H.L., 2020. Clinical landscape of oncolytic virus research in 2020. J. Immunother. Cancer 8 (2).

Maeda, K., Akira, S., 2017. Regulation of mRNA stability by CCCH-type zinc-finger proteins in immune cells. Int. Immunol. 29 (4), 149–155.

Maeda, H., Wu, J., Sawa, T., Matsumura, Y., Hori, K., 2000. Tumor vascular permeability and the EPR effect in macromolecular therapeutics: a review. J. Control. Release 65 (1–2), 271–284.

Manning, B.D., Cantley, L.C., 2007. AKT/PKB signaling: navigating downstream. Cell 129 (7), 1261–1274.

Mansurov, A., Ishihara, J., Hosseinchi, P., Potin, L., Marchell, T.M., Ishihara, A., et al., 2020. Collagen-binding IL-12 enhances tumour inflammation and drives the complete remission of established immunologically cold mouse tumours. Nat. Biomed. Eng. 4 (5), 531–543.

Martin, W.L., West Jr., A.P., Gan, L., Bjorkman, P.J., 2001. Crystal structure at 2.8 a of an FcRn/heterodimeric fc complex: mechanism of pH-dependent binding. Mol. Cell 7 (4), 867–877.

Medina-Echeverz, J., Fioravanti, J., Diaz-Valdes, N., Frank, K., Aranda, F., Gomar, C., et al., 2014. Harnessing high density lipoproteins to block transforming growth factor beta and to inhibit the growth of liver tumor metastases. PLoS One 9 (5), e96799.

Minute, L., Teijeira, A., Sanchez-Paulete, A.R., Ochoa, M.C., Alvarez, M., Otano, I., et al., 2020. Cellular cytotoxicity is a form of immunogenic cell death. J. Immunother. Cancer 8 (1).

Mo, F., Yu, Z., Li, P., Oh, J., Spolski, R., Zhao, L., et al., 2021. An engineered IL-2 partial agonist promotes CD8(+) T cell stemness. Nature 597 (7877), 544–548.

Momin, N., Mehta, N.K., Bennett, N.R., Ma, L., Palmeri, J.R., Chinn, M.M., et al., 2019. Anchoring of intratumorally administered cytokines to collagen safely potentiates systemic cancer immunotherapy. Sci. Transl. Med. 11 (498).

Mortier, E., Quemener, A., Vusio, P., Lorenzen, I., Boublik, Y., Grotzinger, J., et al., 2006. Soluble interleukin-15 receptor alpha (IL-15R alpha)-sushi as a selective and potent agonist of IL-15 action through IL-15R beta/gamma. Hyperagonist IL-15 x IL-15R alpha fusion proteins. J. Biol. Chem. 281 (3), 1612–1619.

Mullen, L., Rigby, A., Sclanders, M., Adams, G., Mittal, G., Colston, J., et al., 2014. Latency can be conferred to a variety of cytokines by fusion with latency-associated peptide from TGF-beta. Expert Opin. Drug Deliv. 11 (1), 5–16.

Mumm, J.B., Emmerich, J., Zhang, X., Chan, I., Wu, L., Mauze, S., et al., 2011. IL-10 elicits IFNgamma-dependent tumor immune surveillance. Cancer Cell 20 (6), 781–796.
Murer, P., Neri, D., 2019. Antibody-cytokine fusion proteins: a novel class of biopharmaceuticals for the therapy of cancer and of chronic inflammation. Nat. Biotechnol. 52, 42–53.
Naing, A., Papadopoulos, K.P., Autio, K.A., Ott, P.A., Patel, M.R., Wong, D.J., et al., 2016. Safety, antitumor activity, and immune activation of Pegylated recombinant human Interleukin-10 (AM0010) in patients with advanced solid tumors. J. Clin. Oncol. 34 (29), 3562–3569.
Naing, A., Wong, D.J., Infante, J.R., Korn, W.M., Aljumaily, R., Papadopoulos, K.P., et al., 2019. Pegilodecakin combined with pembrolizumab or nivolumab for patients with advanced solid tumours (IVY): a multicentre, multicohort, open-label, phase 1b trial. Lancet Oncol. 20 (11), 1544–1555.
Negrier, S., Escudier, B., Lasset, C., Douillard, J.Y., Savary, J., Chevreau, C., et al., 1998. Recombinant human interleukin-2, recombinant human interferon alfa-2a, or both in metastatic renal-cell carcinoma. Groupe Francais d'Immunotherapie. N. Engl. J. Med. 338 (18), 1272–1278.
Nischan, N., Hackenberger, C.P., 2014. Site-specific PEGylation of proteins: recent developments. J. Organomet. Chem. 79 (22), 10727–10733.
Ober, R.J., Martinez, C., Vaccaro, C., Zhou, J., Ward, E.S., 2004. Visualizing the site and dynamics of IgG salvage by the MHC class I-related receptor FcRn. J. Immunol. 172 (4), 2021–2029.
Ochoa, M.C., Fioravanti, J., Rodriguez, I., Hervas-Stubbs, S., Azpilikueta, A., Mazzolini, G., et al., 2013. Antitumor immunotherapeutic and toxic properties of an HDL-conjugated chimeric IL-15 fusion protein. Cancer Res. 73 (1), 139–149.
Ochoa, M.C., Minute, L., Lopez, A., Perez-Ruiz, E., Gomar, C., Vasquez, M., et al., 2018. Enhancement of antibody-dependent cellular cytotoxicity of cetuximab by a chimeric protein encompassing interleukin-15. Onco. Targets. Ther. 7 (2), e1393597.
Olsen, S.K., Ota, N., Kishishita, S., Kukimoto-Niino, M., Murayama, K., Uchiyama, H., et al., 2007. Crystal structure of the interleukin-15. Interleukin-15 receptor alpha complex: insights into trans and cis presentation. J. Biol. Chem. 282 (51), 37191–37204.
Opal, S.M., DePalo, V.A., 2000. Anti-inflammatory cytokines. Chest 117 (4), 1162–1172.
O'Shea, J.J., Schwartz, D.M., Villarino, A.V., Gadina, M., McInnes, I.B., Laurence, A., 2015. The JAK-STAT pathway: impact on human disease and therapeutic intervention. Annu. Rev. Med. 66, 311–328.
Pastor, F., Berraondo, P., Etxeberria, I., Frederick, J., Sahin, U., Gilboa, E., et al., 2018. An RNA toolbox for cancer immunotherapy. Nat. Rev. Drug Discov. 17 (10), 751–767.
Patel, M.R., Bauer, T.M., Jimeno, A., Wang, D., LoRusso, P., Do, K.T., et al., 2020. A phase I study of mRNA-2752, a lipid nanoparticle encapsulating mRNAs encoding human OX40L, IL-23, and IL-36γ, for intratumoral (iTu) injection alone and in combination with durvalumab. J. Clin. Oncol. 38 (suppl. 15), 3092.
Pegram, H.J., Lee, J.C., Hayman, E.G., Imperato, G.H., Tedder, T.F., Sadelain, M., et al., 2012. Tumor-targeted T cells modified to secrete IL-12 eradicate systemic tumors without need for prior conditioning. Blood 119 (18), 4133–4141.
Perez-Ruiz, E., Minute, L., Otano, I., Alvarez, M., Ochoa, M.C., Belsue, V., et al., 2019. Prophylactic TNF blockade uncouples efficacy and toxicity in dual CTLA-4 and PD-1 immunotherapy. Nature 569 (7756), 428–432.
Propper, D.J., Balkwill, F.R., 2022. Harnessing cytokines and chemokines for cancer therapy. Nat. Rev. Clin. Oncol. 19, 237–253.
Ricard-Blum, S., 2011. The collagen family. Cold Spring Harb. Perspect. Biol. 3 (1), a004978.

Robertson, M.J., Kirkwood, J.M., Logan, T.F., Koch, K.M., Kathman, S., Kirby, L.C., et al., 2008. A dose-escalation study of recombinant human interleukin-18 using two different schedules of administration in patients with cancer. Clin. Cancer Res. 14 (11), 3462–3469.

Rodriguez-Ruiz, M.E., Rodriguez, I., Mayorga, L., Labiano, T., Barbes, B., Etxeberria, I., et al., 2019. TGFbeta blockade enhances radiotherapy Abscopal efficacy effects in combination with anti-PD1 and anti-CD137 Immunostimulatory monoclonal antibodies. Mol. Cancer Ther. 18 (3), 621–631.

Rosenberg, S.A., Restifo, N.P., Yang, J.C., Morgan, R.A., Dudley, M.E., 2008. Adoptive cell transfer: a clinical path to effective cancer immunotherapy. Nat. Rev. Cancer 8 (4), 299–308.

Rossey, I., Gilman, M.S., Kabeche, S.C., Sedeyn, K., Wrapp, D., Kanekiyo, M., et al., 2017. Potent single-domain antibodies that arrest respiratory syncytial virus fusion protein in its prefusion state. Nat. Commun. 8, 14158.

Russell, S.J., Peng, K.W., Bell, J.C., 2012. Oncolytic virotherapy. Nat. Biotechnol. 30 (7), 658–670.

Sahin, U., Kariko, K., Tureci, O., 2014. mRNA-based therapeutics—developing a new class of drugs. Nat. Rev. Drug Discov. 13 (10), 759–780.

Sanchez-Paulete, A.R., Teijeira, A., Quetglas, J.I., Rodriguez-Ruiz, M.E., Sanchez-Arraez, A., Labiano, S., et al., 2018. Intratumoral immunotherapy with XCL1 and sFlt3L encoded in recombinant Semliki Forest virus-derived vectors fosters dendritic cell-mediated T-cell cross-priming. Cancer Res. 78 (23), 6643–6654.

Sandau, M.M., Schluns, K.S., Lefrancois, L., Jameson, S.C., 2004. Cutting edge: transpresentation of IL-15 by bone marrow-derived cells necessitates expression of IL-15 and IL-15R alpha by the same cells. J. Immunol. 173 (11), 6537–6541.

Sanz, L., Alvarez-Vallina, L., 2021. Engineered mRNA and the rise of next-generation antibodies. Antibodies (Basel) 10 (4).

Sanz, L., Cuesta, A.M., Compte, M., Alvarez-Vallina, L., 2005. Antibody engineering: facing new challenges in cancer therapy. Acta Pharmacol. Sin. 26 (6), 641–648.

Shuai, K., Liu, B., 2003. Regulation of JAK-STAT signalling in the immune system. Nat. Rev. Immunol. 3 (11), 900–911.

Silva, D.A., Yu, S., Ulge, U.Y., Spangler, J.B., Jude, K.M., Labao-Almeida, C., et al., 2019. De novo design of potent and selective mimics of IL-2 and IL-15. Nature 565 (7738), 186–191.

Skrombolas, D., Frelinger, J.G., 2014. Challenges and developing solutions for increasing the benefits of IL-2 treatment in tumor therapy. Expert. Rev. Clin. Immunol. 10 (2), 207–217.

Skrombolas, D., Sullivan, M., Frelinger, J.G., 2019. Development of an Interleukin-12 fusion protein that is activated by cleavage with matrix metalloproteinase 9. J. Interf. Cytokine Res. 39 (4), 233–245.

Sockolosky, J.T., Trotta, E., Parisi, G., Picton, L., Su, L.L., Le, A.C., et al., 2018. Selective targeting of engineered T cells using orthogonal IL-2 cytokine-receptor complexes. Science 359 (6379), 1037–1042.

Solal-Celigny, P., Lepage, E., Brousse, N., Reyes, F., Haioun, C., Leporrier, M., et al., 1993. Recombinant interferon alfa-2b combined with a regimen containing doxorubicin in patients with advanced follicular lymphoma. Groupe d'Etude des Lymphomes de l'Adulte. N. Engl. J. Med. 329 (22), 1608–1614.

Spolski, R., Li, P., Leonard, W.J., 2018. Biology and regulation of IL-2: from molecular mechanisms to human therapy. Nat. Rev. Immunol. 18 (10), 648–659.

Stoklasek, T.A., Schluns, K.S., Lefrancois, L., 2006. Combined IL-15/IL-15Ralpha immunotherapy maximizes IL-15 activity in vivo. J. Immunol. 177 (9), 6072–6080.

Strohl, W.R., 2015. Fusion proteins for half-life extension of biologics as a strategy to make biobetters. BioDrugs 29 (4), 215–239.

Tan, C., Waldmann, T.A., 2020. Bench-to-bedside translation of interleukin-15 for immunotherapy: principles and challenges. Expert Opin. Drug Deliv. 17 (7), 895–898.

Tomasello, E., Pollet, E., Vu Manh, T.P., Uze, G., Dalod, M., 2014. Harnessing mechanistic knowledge on beneficial versus deleterious IFN-I effects to design innovative immunotherapies targeting cytokine activity to specific cell types. Front. Immunol. 5, 526.

Turunen, S.P., Tatti-Bugaeva, O., Lehti, K., 2017. Membrane-type matrix metalloproteases as diverse effectors of cancer progression. Biochim. Biophys. Acta, Mol. Cell Res. 1864 (11 Pt A), 1974–1988.

Tzeng, A., Kwan, B.H., Opel, C.F., Navaratna, T., Wittrup, K.D., 2015. Antigen specificity can be irrelevant to immunocytokine efficacy and biodistribution. Proc. Natl. Acad. Sci. U. S. A. 112 (11), 3320–3325.

Uze, G., Tavernier, J., 2015. High efficiency targeting of IFN-alpha activity: possible applications in fighting tumours and infections. Cytokine Growth Factor Rev. 26 (2), 179–182.

Vanrell, L., Di Scala, M., Blanco, L., Otano, I., Gil-Farina, I., Baldim, V., et al., 2011. Development of a liver-specific Tet-on inducible system for AAV vectors and its application in the treatment of liver cancer. Mol. Ther. 19 (7), 1245–1253.

Vasquez, M., Fioravanti, J., Aranda, F., Paredes, V., Gomar, C., Ardaiz, N., et al., 2016. Interferon alpha bioactivity critically depends on scavenger receptor class B type I function. Onco. Targets. Ther. 5 (8), e1196309.

Vasquez, M., Paredes-Cervantes, V., Aranda, F., Ardaiz, N., Gomar, C., Berraondo, P., 2017a. Antitumor effect of an adeno-associated virus expressing apolipoprotein A-1 fused to interferon alpha in an interferon alpha-resistant murine tumor model. Oncotarget 8 (3), 5247–5255.

Vasquez, M., Simoes, I., Consuegra-Fernandez, M., Aranda, F., Lozano, F., Berraondo, P., 2017b. Exploiting scavenger receptors in cancer immunotherapy: lessons from CD5 and SR-B1. Eur. J. Immunol. 47 (7), 1108–1118.

Vazquez-Lombardi, R., Loetsch, C., Zinkl, D., Jackson, J., Schofield, P., Deenick, E.K., et al., 2017. Potent antitumour activity of interleukin-2-fc fusion proteins requires fc-mediated depletion of regulatory T-cells. Nat. Commun. 8, 15373.

Vinay, D.S., Ryan, E.P., Pawelec, G., Talib, W.H., Stagg, J., Elkord, E., et al., 2015. Immune evasion in cancer: mechanistic basis and therapeutic strategies. Semin. Cancer Biol. 35 (Suppl), S185–S198.

Wang, Z., Guo, J., Sun, J., Liang, P., Wei, Y., Deng, X., et al., 2019a. Thermoresponsive and protease-cleavable interferon-polypeptide conjugates with spatiotemporally programmed two-step release kinetics for tumor therapy. Adv. Sci. (Weinh). 6 (16), 1900586.

Wang, H., Hou, Y., Hu, Y., Dou, J., Shen, Y., Wang, Y., et al., 2019b. Enzyme-activatable interferon-poly(alpha-amino acid) conjugates for tumor microenvironment potentiation. Biomacromolecules 20 (8), 3000–3008.

Xu, A., Bhanumathy, K.K., Wu, J., Ye, Z., Freywald, A., Leary, S.C., et al., 2016. IL-15 signaling promotes adoptive effector T-cell survival and memory formation in irradiation-induced lymphopenia. Cell Biosci. 6, 30.

Xue, D., Moon, B., Liao, J., Guo, J., Zou, Z., Han, Y., et al., 2022. A tumor-specific pro-IL-12 activates preexisting cytotoxic T cells to control established tumors. Sci. Immunol. 7 (67), eabi6899.

Yang, Z.Z., Grote, D.M., Ziesmer, S.C., Xiu, B., Yates, N.R., Secreto, F.J., et al., 2013. Soluble and membrane-bound TGF-beta-mediated regulation of intratumoral T cell differentiation and function in B-cell non-Hodgkin lymphoma. PLoS One 8 (3), e59456.

Yeku, O.O., Purdon, T.J., Koneru, M., Spriggs, D., Brentjens, R.J., 2017. Armored CAR T cells enhance antitumor efficacy and overcome the tumor microenvironment. Sci. Rep. 7 (1), 10541.

Zhang, L., Morgan, R.A., Beane, J.D., Zheng, Z., Dudley, M.E., Kassim, S.H., et al., 2015. Tumor-infiltrating lymphocytes genetically engineered with an inducible gene encoding interleukin-12 for the immunotherapy of metastatic melanoma. Clin. Cancer Res. 21 (10), 2278–2288.

Zhang, Q., Hresko, M.E., Picton, L.K., Su, L., Hollander, M.J., Nunez-Cruz, S., et al., 2021. A human orthogonal IL-2 and IL-2Rbeta system enhances CAR T cell expansion and antitumor activity in a murine model of leukemia. Sci. Transl. Med. 13 (625), eabg6986.

Zhou, Z.H., Ji, C.D., Xiao, H.L., Zhao, H.B., Cui, Y.H., Bian, X.W., 2017. Reorganized collagen in the tumor microenvironment of gastric cancer and its association with prognosis. J. Cancer 8 (8), 1466–1476.

Zhou, T., Damsky, W., Weizman, O.E., McGeary, M.K., Hartmann, K.P., Rosen, C.E., et al., 2020. IL-18BP is a secreted immune checkpoint and barrier to IL-18 immunotherapy. Nature 583 (7817), 609–614.

Zhu, X., Marcus, W.D., Xu, W., Lee, H.I., Han, K., Egan, J.O., et al., 2009. Novel human interleukin-15 agonists. J. Immunol. 183 (6), 3598–3607.

CHAPTER SIX

Emerging applications of nanobodies in cancer therapy

Robin Maximilian Awad[a,†], Fien Meeus[a,†], Hannelore Ceuppens[a], Thomas Ertveldt[a], Heleen Hanssens[b], Quentin Lecocq[a], Lukasz Mateusiak[b], Katty Zeven[b], Hana Valenta[c], Timo W.M. De Groof[b], Yannick De Vlaeminck[a], Ahmet Krasniqi[b], Kim De Veirman[d], Cleo Goyvaerts[a], Matthias D'Huyvetter[b], Sophie Hernot[b], Nick Devoogdt[b], and Karine Breckpot[a,*]

[a]Laboratory for Molecular and Cellular Therapy, Department of Biomedical Sciences, Vrije Universiteit Brussel, Brussels, Belgium
[b]In Vivo Cellular and Molecular Imaging Laboratory, Department of Medical Imaging, Vrije Universiteit Brussel, Brussels, Belgium
[c]Lab for Nanobiology, Department of Chemistry, KU Leuven, Leuven, Belgium
[d]Laboratory for Hematology and Immunology, Department of Biomedical Sciences, Vrije Universiteit Brussel, Brussels, Belgium
*Corresponding author: e-mail address: karine.breckpot@vub.be

Contents

† Contributed equally.

International Review of Cell and Molecular Biology, Volume 369
ISSN 1937-6448
https://doi.org/10.1016/bs.ircmb.2022.03.010

Abstract

Cancer is a heterogeneous disease, requiring treatment tailored to the unique phenotype of the patient's tumor. Monoclonal antibodies (mAbs) and variants thereof have enabled targeted therapies to selectively target cancer cells. Cancer cell-specific mAbs have been used for image-guided surgery and targeted delivery of radionuclides or toxic agents, improving classical treatment strategies. Cancer cell-specific mAbs can further inhibit tumor cell growth or can stimulate immune-mediated destruction of cancer cells, a feature that has also been achieved through mAb-mediated manipulation of immune cells and pathways. Drawbacks of mAbs and their variants, together with the discovery of camelid heavy chain-only antibodies and the many advantageous features of their variable domains, referred to as V_HHs, single domain antibodies or nanobodies (Nbs), resulted in the exploration of Nbs as an alternative targeting moiety. We therefore review the state-of-the-art as well as novel exploitation strategies of Nbs for targeted cancer therapy.

1. Introduction

In the past, a cancer diagnosis inevitably implied death. Yet, since the beginning of the 20th century, advances in surgery, and the emergence of radiation therapy and chemotherapy, enabled the treatment of cancer. Although these modalities significantly improve patient survival, their efficiency is limited. Surgical resection and radiation therapy are restricted to the site of detected tumor lesions and cannot target metastatic niches. Contrarily, chemotherapy acts systemically, but results in collateral damage and serious side effects. To date, several strategies to manipulate the immune system, enabling it to recognize and kill cancer cells, have been developed. Moreover, additional targeted therapies as cancer treatment have been explored. These targeted therapies combine toxic groups or radionuclides with a targeting moiety, thereby restricting the toxic effects to tumor lesions and/or improving the therapy efficiency (Bannas et al., 2017).

Antibodies have been invaluable in the design of immunotherapy and other targeted therapy strategies. Antibodies are large Y-shaped proteins (~150 kDa) composed of two identical heavy and two identical light chains that are bound to each other *via* disulfide bonds and non-covalent interactions (Fig. 1). Three complementarity determining regions (CDRs), intervals of highly diverse sequences, are present in the variable regions of both the heavy (V_H) and light (V_L) chains. The CDR3 region is the main determinant of antigen specificity (Jovčevska and Muyldermans, 2020). The V_H and V_L chains are held together by hydrophobic interactions within their

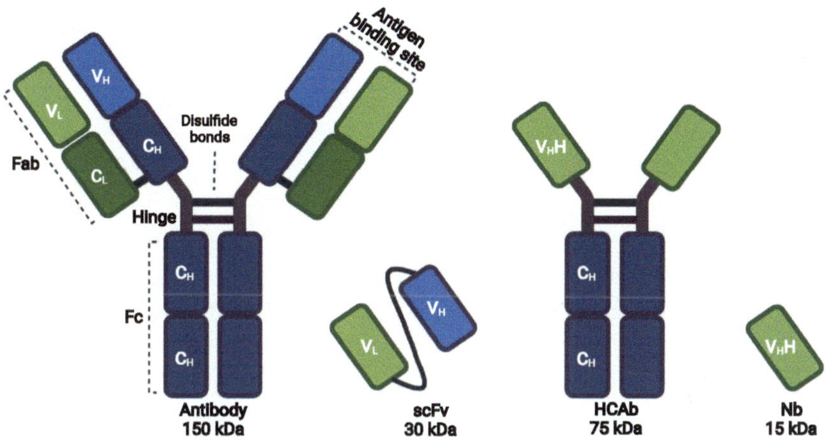

Fig. 1 Graphical representation of a conventional mAb, HCAb and derived antibody fragments. Abbreviations: Fab, antigen binding fragment; Fc, fragment crystallizable or constant region; HCAb, heavy chain-only antibody; Nb, nanobody; scFv, fusion protein of the V_H and V_L chains of antibodies.

framework regions and are stabilized through their linkage to constant domains that are associated by disulfide bridges. Together these form the antigen-binding fragment (Fab) region, which is coupled to a fragment crystallizable or constant region (Fc) *via* a disulfide bond-containing hinge (Inbar et al., 1972; Pauling, 1940; Porter, 1959). The Fc region can engage Fc receptors on immune cells or complement component 1q (C1q), a key molecule in the complement system, and as such can mediate effector functions, such as antibody-dependent cell-mediated cytotoxicity (ADCC), antibody-dependent cell-mediated phagocytosis (ADCP) and complement-dependent cytotoxicity (CDC) (Gul et al., 2014; Moeller, 1965; Rouard et al., 1997; Zahavi and Weiner, 2020). Over the years, antibodies, in particular monoclonal antibodies (mAbs), have been used in cancer therapy for various reasons: to deliver cytotoxic agents to cancer cells (antibody-drug conjugates, ADCs), to deprive cancer cells of survival signals, for instance neutralizing receptor-ligand interactions, to mark cancer cells for destruction by the immune system, to inhibit angiogenesis as well as to support the activity of anticancer immune cells, exemplified by antibodies that target inhibitory immune checkpoints (ICPs). Notwithstanding the successful use of therapeutic mAbs, there are also certain limitations, including high production costs, susceptibility to proteolytic degradation, immunogenicity resulting in antidrug antibodies (ADAs), long blood circulation time, restricted access to epitopes and limited ability to penetrate into tissues such as tumors

(Broos et al., 2019a; Epenetos et al., 1986; Harding et al., 2010). The latter might be due to the size of antibodies, though it has been suggested that high affinity antibodies bind to the first antigen they encounter and therefore are retained, prohibiting further diffusion (Juweid et al., 1992). Fusion proteins of the V_H and V_L chains of mAbs, referred to as single-chain variable fragments (scFvs) (~30 kDa) (Fig. 1), have been used in cancer therapy as these are considerably smaller than mAbs. However, the artificial nature of scFvs comes at the cost of solubility, stability and aggregation issues (Bannas et al., 2017; Hanssens et al., 2021; Verhaar et al., 2021). Nonetheless, scFvs allowed the design of novel mAb-inspired tools, including antigen-presenting cell (APC)-targeted viral vectors and cancer cell-targeted chimeric antigen receptors (CARs) (Ageichik et al., 2011; Bao et al., 2021; Gennari et al., 2009; Sievers et al., 2020; Xie et al., 2019).

Camelids contain B-cell populations that produce conventional antibodies as well as heavy chain-only antibodies (HCAbs, 75 kDa), which lack light chains together with the first constant domain of the heavy chain (CH1) (Fig. 1) (Hamers-Casterman et al., 1993). Due to the lack of a light chain, paratope diversity depends on three CDRs as opposed to six CDRs in conventional antibodies. Yet, this is compensated by a usually larger CDR3 region, the most important region in determining antibody specificity and diversity. The prolate shaped variable binding domains of HCAbs, referred to as V_HHs, single domain antibodies or nanobodies (Nbs), are small in size (~15 kDa) and can be produced as recombinant HCAb fragments. These Nbs show affinities equal to mAbs and have several advantages over mAbs and scFvs: relative low cost of production (at least in the preclinical setting), high water solubility and stability, including resistance to proteolytic degradation, access to epitopes that are inaccessible to mAbs, low immunogenicity, fast circulation times, excellent penetration of tumor tissue, low off-target accumulation and sometimes even the ability to cross the blood-brain-barrier (Abulrob et al., 2005; Ackaert et al., 2021; Broos et al., 2019a; Debie et al., 2020; Jovčevska and Muyldermans, 2020). Moreover, the single domain nature facilitates molecular engineering, therefore easy construction of multivalent, multispecific, chimeric HCAbs containing a human Fc-domain and CARs, paving the way for a variety of new and improved cancer therapy applications inspired on applications initially performed with mAbs or variants thereof (Fig. 2).

The applications of Nbs within specific cancer therapeutic approaches will be further discussed in this review.

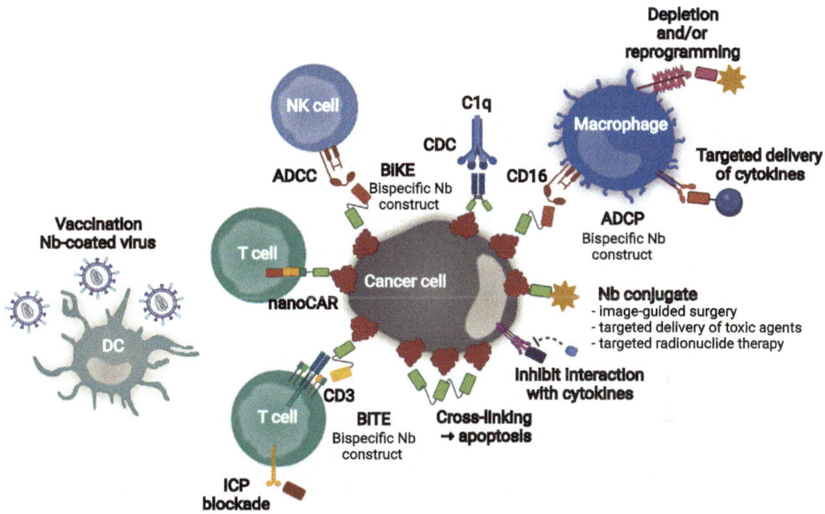

Fig. 2 Graphical representation of cancer therapy strategies in which Nbs are explored. Abbreviations: ADCC, antibody-dependent cell-mediated cytotoxicity; ADCP, antibody-dependent cell-mediated phagocytosis; BiKE, bispecific killer-cell engager; BiTE, bispecific T-cell engager; CDC, complement-dependent cytotoxicity; DC, dendritic cell; ICP, immune checkpoint; Nb, nanobody; nanoCAR, chimeric antigen receptor using a Nb as a targeting moiety and NK, natural killer.

2. Nanobodies: Small antigen-binding proteins with a big impact on conventional cancer therapies

Cancer-specific mAbs have been used to improve conventional cancer therapies by enabling the detection of tumors and by allowing the delivery of toxins and radionuclides in a targeted manner. However, the big size and the Fc tail of mAbs imply a prolonged plasma half-life and therefore diminish contrast of tracers and enhance systemic toxicity of ADCs and targeted radionuclides. In contrast to bulky mAbs, the low molecular weight of Nbs and lack of the Fc region contribute to quick systemic clearance. Another advantage of Nbs for the design of novel targeting agents is their robustness as well as reproducible and uncomplicated conjugation chemistry. Random labeling strategies leading to heterogeneous products with varying labeling degree are frequently employed. Alternatively, Nbs can be labeled site-specifically, giving rise to a homogeneous group of targeting moieties, often through maleimide-sulfhydryl reaction on a cysteine-tag or a sortase-A-mediated strategy (Debie et al., 2017; Massa et al., 2014, 2016; Pleiner et al., 2015). Nbs labeled with fluorescent compounds, toxic agents or radionuclides are then used for the detection or eradication of tumor cells, which will be discussed in the following paragraphs.

2.1 Nanobody-mediated image-guided surgery

Surgery remains the basis of oncological treatment for many solid cancers diagnosed at an early stage but can also be applied with the intention of reducing the tumor volume *via* debulking, as a prelude to chemo- or radiotherapy. Although in an ideal situation the tumor is completely eradicated without leaving any cancer cell behind, incomplete excision and positive surgical margins are frequently reported. This is a major factor contributing to the high recurrence rate of cancer and has thus strong prognostic implications (Hinni et al., 2013; Lam et al., 2008; O'Kelly Priddy et al., 2016; Orosco et al., 2018; Yossepowitch et al., 2014).

Various diagnostic modalities have been developed to guide surgeons during surgery to localize and delineate tumors. Development of these modalities grew into the booming field of image-guided surgery.

Nonetheless, modalities such as X-ray computed tomography (CT), magnetic resonance imaging or ultrasound fall short in terms of real-time intraoperative applications, bulkiness, specificity or not being contactless. Considerable focus has been recently directed toward fluorescence-guided surgery (FGS) as an overall cheaper, highly sensitive and easy to implement real-time technology based on the detection of light emitted by an excited fluorophore *in situ* (Hernot et al., 2019; Nagaya et al., 2017). For surgical applications, fluorescent contrast agents characterized by photon emission in the far-red and near-infrared range (NIR) (650–900 nm) are mostly studied as these provide considerable advantages in terms of reduced autofluorescence and allow up to several millimeters tissue penetration. Besides the non-specific fluorescent dye indocyanine green (ICG) and the prodrug 5-aminolevulinic acid (5-ALA) being routinely used for the visualization of respectively liver tumors and high-grade glioma's, OTL-38, a fluorescent-derivative of folic acid recently got approved by the FDA under the name Cytalux as the first tumor-targeted fluorescent agent for ovarian cancer. Many other targeted fluorescent tracers, often based on mAbs, are currently undergoing clinical evaluation (Debie and Hernot, 2019; Hernot et al., 2019; Lauwerends et al., 2021).

Nb-based contrast agents are currently being explored in a preclinical setting for FGS due to their favorable pharmacokinetic parameters (Table 1). As demonstrated by Oliveira (2012) and Kijanka (2013) in subcutaneous tumor models, the quick systemic clearance of fluorescent Nbs translates to a significantly shorter time interval between injection and highly specific imaging with sufficient tumor contrast, as compared to their

Table 1 Nb-based tracers for image-guided surgery.

Target	Cancer model	Compound and dye	Specificity	Status	Reference
CEA	Pancreas and colon cancer	Nb CEA5	Human	Preclinical	Lwin et al. (2020)
		IRDye800CW			Lwin et al. (2021)
CAIX	Breast cancer	B9 IRDye800CW	Human	Preclinical	van Brussel et al. (2016)
EGFR	Squamous carcinoma	7D12	Human	Preclinical	Oliveira et al. (2012)
		IRDye800CW, IRDye680RD			van Driel et al. (2016)
		7D12 ^{111}In/IRDye700DX	Human	Preclinical	Renard et al. (2021)
HER2	Breast cancer	11A4	Human	Preclinical	Kijanka et al. (2013, 2016)
		IRDye800CW			
		2Rs15d	Human	Preclinical	Debie et al. (2017, 2018)
		IRDye800CW, IRDye680RD			
		2Rs15d Cy5/^{111}In	Human	Preclinical	Debie et al. (2021)
CAIX + HER2	Breast cancer	B9, 11A4 IRDye800CW	Human	Preclinical	Kijanka et al. (2016)

Abbreviations: CAIX, carbonic anhydrase IX; CEA, carcinoembryonic antigen; EGFR, epidermal growth factor receptor; HER2, human epidermal growth factor receptor 2.

antibody counterparts (30 min instead of 24 h). Consequently, Nbs labeled with the widely used IRDye800CW, have been evaluated in more advanced animal models, providing further evidence for the feasibility and impact of image-guided surgery using Nbs as a targeting moiety. For example, the Nbs 7D12 and B9, targeting respectively the epidermal growth factor receptor-1 (EGFR) or carbonic anhydrase IX (CAIX) were successfully used to delineate and resect orthotopically implanted tongue tumors or ductal carcinoma *in situ* cells. Derived lymph node metastases were detectable as well (van Brussel et al., 2016; van Driel et al., 2014). Likewise, Debie (2018) showed that dispersed ovarian cancer nodules within the abdomen of mice could rapidly and specifically be visualized using Nb 2Rs15d that targets human epidermal growth factor receptor 2 (HER2). This significantly improved peritoneal tumor debulking in comparison to conventional surgical strategies. Recently, Lwin (2020, 2021) used Nb CEA5 specific for carcinoembryonic antigen (CEA) to highlight orthotopic pancreatic and colon cancer (satellite) lesions, showing that these were detectable with FDA-approved fluorescent cameras, including that of the da Vinci surgical robot. The increased tumor-to-background ratios obtained with simultaneous targeting of more than one biomarker using fluorescent Nbs (*e.g.*, HER2 [11A4] in combination with CAIX [B9]) suggests an added value of using a mixture of Nbs targeting different receptors and emitting light at different wavelengths for the detection of heterogeneous tumors (Kijanka et al., 2016), however, clinical implementation might be complicated. Nb-based FGS has also been explored in combination with radioactivity (Renard et al., 2021). Combining FGS with a nuclear component facilitates the navigation toward fluorescent signals in deeply located tissues *via* gamma probe tracing. In addition, the radioactive component can be readily used for better surgical planning *via* preoperative imaging (van Leeuwen et al., 2016). In Debie (2021), the anti-HER2 Nb 2Rs15d was modified with a fluorescent dye and a radioisotope, both incorporated into a single label to overcome tracer heterogeneity. Intraperitoneal ovarian cancer lesions could then successfully be imaged *via* both single photon emission computed tomography (SPECT)/CT and intraoperative fluorescence imaging (Debie et al., 2021).

2.2 Nanobody-mediated delivery of toxic agents to cancer cells

Several Nbs that target tumor markers key in tumor promoting signaling cascades have been developed. These include Nbs targeting EGFR1/2, hepatocyte growth factor (HGF), c-mesenchymal-epithelial transition factor (c-Met), vascular endothelial growth factor receptor 2 (VEGFR2) and C-X-C chemokine receptor type 7 (CXCR7). As many Nbs lack an inherent therapeutic effect, conjugation with toxic or effector molecules is necessary (Hu et al., 2017). Various Nb-based delivery strategies for targeted cancer therapy have been explored such as drug loaded nanoparticles (NPs) decorated with Nbs, chemical Nb conjugations or genetic fusion of toxic proteins to Nbs (Table 2).

Liposomes, micelles, dendrimers, nanospheres and nanocapsules are among the NPs that can be loaded with drugs and that have been coupled to Nbs (Hu et al., 2017). These targeted NPs release their cargo inside the cell after receptor-mediated endocytosis, thereby reducing damage to healthy tissue (Kang and Jung, 2019; Patra et al., 2018; Yang et al., 2017). Oliveira (2010) were the first to show that PEGylated liposomes decorated with anti-EGFR Nbs (Ega1) were internalized by cancer cells, thereby downregulating EGFR and inhibiting tumor growth. Loading this Ega1 liposome with AG538, a kinase inhibitor targeting insulin growth factor 1 receptor, resulted in even stronger growth inhibition (van der Meel et al., 2012). Recently, anthrax lethal factor containing NPs were conjugated with anti-CD19 Nbs. It has been shown that these conjugates selectively eradicate CD19 positive cells without toxicity to normal cells (Banihashemi et al., 2021). Micelles are aggregates of hydrophilic heads directed toward the aqueous solution with hydrophobic tails directed to the center. This composition allows to carry hydrophobic drugs stably through the bloodstream (Hu et al., 2017). Armoring these micelles with Nbs to generate a targeted delivery system toward cancer cells was explored using anti-EGFR Nbs (Ega1). These EGFR-targeted micelles carrying doxycycline showed significantly improved therapeutic efficacy in cell lines and xenografted tumors compared to non-targeted micelles (Talelli et al., 2011, 2013). Polymeric NPs such as dendrimers, nanospheres and nanocapsules are considered one of the most efficient vehicles for drug delivery and have already been explored for cancer therapy (Hu et al., 2017). Recently, Wu et al. (2020) developed a novel strategy where they linked anti-EGFR Nbs (7D12) to dendrimers. These multivalent Nb drug conjugates (NDCs) could successfully target and deliver a tetravalent platinum prodrug to tumors, thereby inhibiting their growth *in vivo* (Wu et al., 2020). Albumin, the most abundant protein in the bloodstream, is a biocompatible and safe vehicle to deliver therapeutics (Hu et al., 2017; Muller et al., 1996). This insight on the potential applicability of albumin as a safe drug delivery system inspired several groups to develop albumin NPs, and later on, albumin NPs decorated with cancer antigen targeting Nbs

Table 2 Nb-toxic drug conjugates.

Target	Cancer model	Compound and drug	Specificity	Status	Reference
CD19	B cell malignancy	Anti-CD19 Nb-coated NPs loaded with anthrax lethal factor	Human	Preclinical	Banihashemi et al. (2021)
EGFR	Head and neck squamous cell carcinoma	Ega1-kinase inhibitor 17864	Human	Preclinical	Altintas et al. (2013)
		AG538			Oliveira et al. (2010)
		DOX			Talelli et al. (2013)
		Platinum			van der Meel et al. (2013)
EGFR	Glioblastoma	ENb1/2 + TRAIL	Human	Preclinical	van de Water et al. (2012)
EGFR	Brain metastatic breast cancer	E_V Death receptor ligand	Human Murine	Preclinical	Kitamura et al. (2021)
EGFR	Head and neck cancer	7D12	Human	Preclinical	Heukers et al. (2014)
		7D12-9G8			van Driel et al. (2016)
		IRDye700DX			de Bruijn et al. (2020)
VEGFR2	293KDR	3VGR19 + PE38	Human	Preclinical	Behdani et al. (2013)
CD7	T-ALL	VHH6 + PE38	Human	Preclinical	Yu et al. (2017)

CEA	Colon cancer	Nb CEA5 *Enterobacter cloacae* β-lactamase + prodrug	Human	Preclinical	Cortez-Retamozo et al. (2004)
HER2	Breast cancer	VHH-polyamidoamine dendrimers with PE38 pDNA	Human	Preclinical	Moradian and Rahbarizadeh (2021)
HER2	Breast cancer	1D5 1D5-18A12 IRDye700DX	Human	Preclinical	Deken et al. (2020)
HGFR	Ovarian cancer, squamous carcinoma, lung cancer, gastric adenocarcinoma	G2	Human	Preclinical	Heukers et al. (2014)
GPCR US28	Glioma	VUN100 IRDye700DX	Human	Preclinical	De Groof et al. (2019)

Abbreviations: CEA, carcinoembryonic antigen; DOX, doxycycline; EGFR, epidermal growth factor receptor; GPCR, G-protein coupled receptor; HGFR, hepatocyte growth factor receptor; NP, nanoparticle; pDNA, plasmid DNA; PDT, photodynamic therapy; PE38, Pseudomonas exotoxin A; T-ALL, T-cell acute lymphoblastic leukemia; VEGFR, vascular endothelial growth factor receptor.

(Altintas et al., 2013; Elzoghby et al., 2012; Hu et al., 2017). In comparison to PEGylated NPs, PEGylated NPs targeting EGFR (Ega1) *via* Nbs and loaded with a kinase inhibitor showed a significantly higher binding to EGFR positive target cells leading to successful internalization, intracellular release of the drug, and antiproliferative effects (Altintas et al., 2013; Hu et al., 2017).

Encapsulation of drugs in NPs is promising but entails several challenges in terms of production and chemistry. ADCs consist of mAbs that are linked to a toxic or effector group. In contrast to NPs, ADCs are more straightforward to produce. Currently over 80 ADCs are under clinical development and 9 are approved by the FDA (Joubert et al., 2020). However, ADCs are large in size, which hampers their migration into the dense tumor bed. NDCs on the other hand combine the advantage of deep tissue penetration, high specificity and a straightforward production process. Conjugation of Nbs to toxic drugs or proteins can be done either by chemical conjugation or genetic fusion (Hu et al., 2017). Although proven successful, chemical conjugation must be approached with caution, as the conjugation method can introduce variability between batches, drug-to-antibody/Nb ratios and/or steric effects, thereby negatively impacting on the efficacy of the resulting drug (Hoffmann et al., 2018; Lucas et al., 2018). Gene-based delivery of a therapeutic on the other hand obviates the need for those cumbersome conjugation and generation procedures yet requires that the toxic moiety is a protein. Several groups have already explored the therapeutic potential of gene-derived NDCs. For example, anti-EGFR Nbs (E_V, ENb1 and ENb2) linked to tumor necrosis factor (TNF)-related apoptosis-inducing ligand (TRAIL) produced by genetically modified stem cells demonstrated therapeutic efficacy in diverse cancer models, among which breast cancer and glioblastoma models (Kitamura et al., 2021; van de Water et al., 2012). The receptor of TRAIL, death receptor 5, has been explored as a therapeutic target for Nb cancer therapy earlier. It has been shown that multivalent Nbs can mimic the natural function of TRAIL and drive apoptosis more efficiently than conventional antibodies and monomeric Nbs (Huet et al., 2014; Sadeghnezhad et al., 2019). Tumor cells that express VEGFR2 could be successfully killed by VEGFR2-specific Nbs (3VGR19) coupled to an active (truncated) variant of Pseudomonas exotoxin A (PE38) (Behdani et al., 2013). A highly active (mutated) variant of PE38 was genetically fused to CD7 targeting Nbs (VHH6) as well. *In vitro*, this NDC exhibited strong antitumor effects in CD7 positive cell lines and primary T-cell acute lymphoblastic leukemia (T-ALL) cells. In a xenograft model with patient-derived T-ALL cells, survival was increased upon administration of Nb-PE38 fusions (Yu et al., 2017). Besides for direct delivery of PE38, Nbs have been recently used for gene mediated PE38 delivery. Polyamidoamine dendrimers were conjugated with Nbs targeting HER2 and were used for the delivery of PE38 encoding plasmid DNA. Upon delivery, exclusively HER2 positive cells expressed PE38 resulting in selective induction of cell death (Moradian and Rahbarizadeh, 2021). Also, genetically fused NDCs were shown applicable for conversion of prodrugs into their active form at specific locations. An anti-CEA Nb (CEA5) fused to the enzyme β-lactamase efficiently converted the cephalosporin nitrogen mustard prodrug into its active form at the surface of CEA positive colon cancer cells *in vitro*, thereby inducing cytotoxicity (Cortez-Retamozo et al., 2004).

Photodynamic therapy (PDT) is also performed with NDCs that contain tumor-targeting Nbs and photosensitizers, typically IRDye700DX. Once these NDCs bind to tumor cells, the photosensitizers can be activated with NIR light, inducing reactive oxygen species production, that causes necrosis of surrounding cells. A Nb (VUN100) that binds to the G-protein coupled receptor US28, overexpressed on tumors, has been conjugated with IRDye700DX to induce cell death in US28 positive glioma cells (De Groof et al., 2019). Similarly, anti-HER2 Nbs (1D5, 1D5-18A12) have been conjugated to IRDye700DX, enabling significant tumor regression in trastuzumab-resistant HER2 positive tumors upon activation. The therapy was less efficient in tumors with low HER2 expression (Deken et al., 2020). Monovalent (7D12) and biparatopic (7D12-9G8) EGFR targeting Nbs have also been conjugated with IRDye700DX. The NDCs were able to selectively induce tumor cell death in an orthotopic model of head and neck cancer and exhibited low toxicity in healthy tissues (de Bruijn et al., 2020; Heukers et al., 2014; van Driel et al., 2016).

2.3 Nanobody-mediated targeted radionuclide therapy

Targeted radionuclide therapy (TRT) is a systemic treatment that uses radiolabeled cancer cell-specific molecules to deliver a cytotoxic dose of radiation to cancer cells with limited damage to healthy tissues and organs (Gudkov et al., 2015). To achieve this, choosing the appropriate radionuclides with physical characteristics that suit the properties of the targeting molecules is crucial for the successful development of TRT agents. Radionuclides available for TRT are divided in alpha (α) particle, beta minus (β^-) particle and in

Auger-electron emitters (Pouget et al., 2011). An important difference between these types of radiation is the linear energy transfer (LET), the energy that is released by the emitted particles along their track. For the same absorbed dose (in Gray—Gy), the high LET of an α-emitter is more cytotoxic than the low LET of β^- or Auger-electron emitters (Pouget et al., 2011), which can be advantageous for the treatment of radioresistant solid tumors (Kassis and Adelstein, 2005). So far, various targeting molecules have been generated and fine-tuned to suit TRT purposes, such as radiolabeled full-sized mAbs, mAb fragments, small proteins and peptides (Gill et al., 2017). Compared to large radiolabeled mAbs, small molecules have shown remarkable clinical success in treating solid tumors due to the lower toxicity profile and better tumor penetration (Czernin et al., 2021; Marcucci et al., 2013; Scott et al., 2012; Witzig et al., 2003).

Various Nbs have been generated and successfully developed as TRT agents (Table 3). One of the most investigated and clinically advanced radiolabeled Nbs is the HER2-specific Nb 2Rs15d (Vaneycken et al., 2011). Nb 2Rs15d, labeled with β^--particle emitting Lutetium-177 (^{177}Lu) exhibited specific tumor uptake in mice bearing HER2 positive tumors, with very low accumulation in non-target organs, except kidneys. Kidney retention could be significantly reduced by co-injection of Gelofusine, increasing the serum half-life of the Nbs. Dosimetry calculation of ^{177}Lu-2Rs15d co-administrated with Gelofusine revealed an equivalent value of absorbed dose delivered to both tumor and kidneys. A significantly prolonged median survival with almost complete regression of the tumor was observed in mice treated with ^{177}Lu-2Rs15d, without signs of radiotoxicity (D'Huyvetter et al., 2014). In a second study, 2Rs15d was labeled with β^--particle emitting Iodine-131 (^{131}I). In HER2 positive tumor xenografted mice, ^{131}I-2Rs15d demonstrated specific tumor uptake, with extremely low accumulation in healthy organs, with an absorbed dose to the tumor higher than to kidneys. In therapeutic modality, ^{131}I-2Rs15d significantly extended the median survival of treated mice with no signs of radiotoxicity (D'Huyvetter et al., 2017). Based on these promising results, ^{131}I-2Rs15d was recently successfully evaluated in a phase I trial including patients with advanced HER2 positive breast cancer (D'Huyvetter et al., 2021).

In an attempt to further increase the therapeutic potential, the same Nb was labeled with α-emitters Astatine-211 (^{211}At), Bismuth-213 (^{213}Bi), and Actinium-225 (^{225}Ac). In mice bearing HER2 positive tumors, ^{211}At-2Rs15d showed specific tumor targeting with relatively low accumulation in major healthy organs, except kidneys. The high kidney retention of

Table 3 Nb-based TRT compounds.

Target	Cancer model	Compound and isotope	Specificity	Status	Reference
CD20	Melanoma	^{177}Lu-Nb 9079	Human	Preclinical	Krasniqi et al. (2017)
CD38	MM	^{177}Lu-2F8	Human	Preclinical	Duray et al. (2021)
CS1	MM	^{225}Ac-sdAb1	Murine	Preclinical	De Veirman et al. (2021)
HER2	Ovarian adenocarcinoma, brain metastatic breast cancer	^{177}Lu-2Rs15d	Human	Clinical	Vaneycken et al. (2011)
		^{131}I-2Rs15d			D'Huyvetter et al. (2014, 2017, 2021)
		^{221}At-2Rs15d			Dekempeneer et al. (2019, 2020)
		^{213}Bi-2Rs15d			
		^{225}Ac-2Rs15d			Puttemans et al. (2020)
M-protein	MM	^{177}Lu-R3B23	Murine	Preclinical	Lemaire et al. (2014)
		^{177}Lu- and ^{225}Ac-8379			Puttemans et al. (2021)
MMR	Mammary adenocarcinoma	^{177}Lu-Nb MMR	Human	Preclinical	Bolli et al. (2019)

Abbreviations: CS1, CD2 subset 1; HER2, human epidermal growth factor receptor 2; MM, multiple myeloma; MMR, macrophage mannose receptor.

^{211}At-2Rs15d can be considered an important dose-limiting factor for therapy (Dekempeneer et al., 2019). The biodistribution and dosimetry profile of ^{213}Bi-2Rs15d was in line with that of ^{211}At-2Rs15d. Also here, the major dose-limiting organs were kidneys. Despite this, ^{213}Bi-2Rs15d was shown to significantly prolong the survival of mice bearing HER2 positive tumors compared to the control treated group. In addition, co-treatment with the anti-HER2 mAb trastuzumab led to a synergistic effect (Dekempeneer et al., 2020). Administration of ^{225}Ac-2Rs15d revealed specific tumor uptake with low uptake in healthy organs in mice bearing HER2 positive tumors. Co-injection of Gelofusine resulted in a considerable reduction of kidney retention compared to ^{225}Ac-2Rs15d alone (Pruszynski et al., 2018). The same radioconjugate was then used to treat mice bearing HER2 positive brain lesions. Therapy results showed significantly higher median survival in mice that were treated with ^{225}Ac-2Rs15d, in comparison with the control groups, with no significant treatment-related toxicity in major organs such as kidneys and liver (Puttemans et al., 2020).

A human CD20 (hCD20)-targeting Nb (Nb 9079) was also developed as a TRT agent. In mice bearing hCD20 positive B16-cell tumors, ^{177}Lu-Nb 9079 showed specific tumor targeting with very low accumulation in non-target organs, except kidneys. The median survival of mice treated with ^{177}Lu-Nb 9079 was significantly prolonged compared to the control groups (Krasniqi et al., 2017). Nb-based TRT of multiple myeloma (MM), a hematological cancer, has also been evaluated. A Nb (R3B23) specific for the M-protein, a monoclonal protein expressed on the surface of MM cells, was generated and used as a radiolabeled agent for imaging and TRT. In a murine immunocompetent model of MM, mice treated with ^{177}Lu-R3B23 showed 50% lower blood uptake compared to the group of mice treated with ^{177}Lu-labeled non-target Nb, confirming its therapeutic potential for the treatment of MM (Lemaire et al., 2014). Targeting M-protein was further confirmed in a second study where it was shown, in addition to successful Nb-based TRT in syngeneic mice with low 5T33MM load, that the generation and characterization of patient-specific anti-idiotype Nbs was feasible (Puttemans et al., 2021). Other antigens on MM cells have also been targeted by Nb-mediated TRT. An anti-CD38 Nb (2F8) radiolabeled with ^{177}Lu revealed to be potent in a dose-dependent manner in mice bearing CD38 positive tumors (Duray et al., 2021). Targeting of 5T33MM cells was further performed using a Nb (sdAb1) that targets CD2 subset 1 (CS1). TRT using anti-CS1 Nbs radiolabeled with ^{225}Ac significantly prolonged survival of 5T33MM bearing mice. Notably, an increase in CD8 positive T cells coinciding with increased expression of PD-L1 on immune and non-immune cells, implying an interferon-gamma (IFN-γ) signature, was observed. This study suggests that Nb-mediated TRT in addition to specifically destroying cancer cells has the ability to activate the immune system (De Veirman et al., 2021). Yet, the precise mode of action remains to be studied.

A Nb that binds the macrophage mannose receptor (Nb MMR) on tumor-associated macrophages (TAMs), present in the majority of cancers and involved in tumor growth, metastasis, and immune escape, has also been used to develop a TRT agent. In a mouse model of mammary carcinoma, treatment of mice with ^{177}Lu-Nb MMR resulted in significant inhibition of tumor growth compared to the control groups, showing that targeting of tumor stromal cells has high potential as a cancer therapy (Bolli et al., 2019).

3. Nanobodies: Emerging applications in cancer immunotherapy

Cancer immunotherapy plays on the virtue of the immune system to prevent, control and eradicate cancer. Active immunotherapy approaches aim to initiate or further enhance inherent antitumor immune responses. These modalities are often based on vaccination and ICP inhibition. In passive immunotherapy, tumor-specific mAbs, cytokines or adoptive cell transfer is used (Papaioannou et al., 2016). mAbs and scFvs have been applied in the design of improved active and passive immunotherapy strategies. Following their example, Nbs found their way into immunotherapy and are under intensive investigation. In the following paragraphs we will discuss how Nbs are used in cancer immunotherapy applications.

3.1 Nanobodies in the design of cancer vaccines

Tumor-associated antigens (TAAs) enable the host's immune system to distinguish malignant from non-malignant self (van der Bruggen et al., 1991). This led to the concept of cancer vaccination, provoking a TAA-specific immune response to eradicate tumor cells (Vermaelen, 2019). As the most professional APCs, dendritic cells (DCs) are in pole position to prime and stimulate TAA-specific CD8 positive cytotoxic T lymphocytes (CTLs). Hence, the aim of any cancer vaccine is to deliver TAAs and adjuvants to DCs, enabling them to elicit a TAA-specific CTL response (Jou et al., 2021; Paston et al., 2021). Cancer vaccines come in various shapes and sizes ranging from autologous DCs over whole tumor cell lysates, recombinant proteins and peptides to TAA-encoding sequences delivered by DNA, mRNA or viral vectors. DCs represent a multifaceted family of subsets with phenotypic and functional specialties, opening possibilities for their targeted manipulation to tune the immune response (Goyvaerts and Breckpot, 2015; Macri et al., 2016). In fact, targeting of receptors: CD40, CD11c, DEC205, chemokine (C motif) ligand 1 (XCL1) and C-type lectin domain family 9 member A (Clec9A) by means of specific mAbs has been shown to direct peptide and NP-based vaccines toward DCs and promote higher antigen uptake with improved T-cell activation due to increased display of peptides on major histocompatibility complex (MHC) proteins (Bonifaz et al., 2004; Botelho et al., 2019; Cruz et al., 2014; Dhodapkar et al., 2014; Zeng et al., 2018). Likewise, Nbs have been exploited to develop DC-targeted vaccines (Table 4).

We developed the Nb-display technology to target lentiviral vector (LV)-based vaccines to surface markers uniquely expressed on APCs. Specifically, LVs were pseudotyped with a fusogenic but binding defective variant of the vesicular stomatitis virus glycoprotein and an APC-specific Nb. The Nbs studied were Nb DC1.8, specific for immature murine bone marrow-derived DCs; Nb DC2.1, cross-reactive for murine and human APCs and R3.13, specific for human DCs (De Groeve et al., 2010). We showed conventional DC or also plasmacytoid DC and macrophage-specific transduction of human subsets *ex vivo* (Nb R3.13 and DC2.1 resp.) and murine subsets (Nb DC1.8 and DC2.1 resp.) *in vivo* (Goyvaerts et al., 2012, 2013). Nb DC1.8 was also applied to engineer a fiber-fibritin-Nb DC1.8 chimeric molecule to redirect the specificity of an adenoviral vector-based vaccine toward the DC2.4 line *in vitro* with enhanced transgene expression compared to infection with the wild type vector (Sharma et al., 2018). Notably, similar to the report of Cire (2014), who used MHCII-targeted LVs, we showed that intranodal administration of Nb DC2.1-displaying LVs enhanced CD4 positive T-cell proliferation, albeit without functional or therapeutic benefit compared to untargeted LVs

Table 4 Nb-based strategies for targeted delivery of cancer vaccines.

Target	Cancer model	Compound	Specificity	Status	Reference
APC	T cell Lymphoma	Nb (DC1.8, DC2.1, R3_13)-displaying virus (OVA)	Mouse and/or human	Preclinical	Goyvaerts et al. (2012, 2013, 2014) Sharma et al. (2018)
CD11b	HPV positive cancer	VHH_{CD11b} + E7	Murine	Preclinical	Woodham et al. (2018)
MHC-II	Melanoma	DC15 + OVA	Murine	Preclinical	Crowley et al. (2020)

Abbreviations: APC, antigen-presenting cell; E7, antigen of HPV; HPV, human papilloma virus; OVA, ovalbumin.

(Goyvaerts et al., 2014). The latter is explained by the fact that targeted LVs trigger an antiviral type I IFN response, impacting on the TAA expression in APCs (Goyvaerts et al., 2017). In contrast, Nb-based targeting has been shown to improve the efficacy of peptide-based cancer vaccines. Intraperitoneal delivery of an anti-CD11b Nb (VHH_{CD11b}) conjugated to the immunodominant epitope of the human papilloma virus antigen E7 reduced the growth of E7 positive tumors more compared to unconjugated E7 peptide. Non-invasive positron emission tomography (PET) imaging with an ^{89}Zirconium-conjugated anti-CD8 Nb showed that the VHH_{CD11b}-conjugated E7 peptide induced higher numbers of tumor-infiltrating lymphocytes (TILs) that correlated with tumor regression and survival (Woodham et al., 2018). In line with these observations, Crowley et al. (2020) showed that footpad injection of an anti-MHCII Nb (DC15) conjugated to the immuno-dominant epitope of ovalbumin (OVA) resulted in the generation of more OVA-specific CTLs than its unconjugated variant. In contrast, it was shown in mouse models of experimental autoimmune encephalomyelitis, type 1 diabetes and rheumatoid arthritis that systemic administration of a single dose of anti-MHCII Nbs conjugated to self-antigens induced tolerance rather than immunity against these antigens (Pishesha et al., 2021). The opposing results obtained with Nb-displaying LVs and Nb-antigen fusion proteins highlight the need for an inflammatory environment in order to activate CTL responses.

3.2 Nanobodies to improve recognition and killing of cancer cells by cytolytic immune cells

Lysis of cancer cells is mainly attributed to CD8 positive CTLs and natural killer (NK) cells. Although cancer therapy has focused on these cytolytic cell types, other cells have come to the forefront as candidates for tumor cell lysis, among other CD4 positive T helper 1 cells and γδT cells (Bonehill et al., 2005; Coulie et al., 2014; Minetto et al., 2019; Miyashita et al., 2021). Nbs have been explored in the design of cancer therapies that focus on cytolytic immune cells, developing strategies to redirect antigen specificity *via* CARs or to engage activating receptors (*e.g.*, CD3 on T cells *versus* CD16 on NK cells).

3.2.1 NanoCARs as tools to direct cancer cell killing

Adoptive cell therapy (ACT) has gained increasing interest as a cancer immunotherapy. TIL or T-cell receptor (TCR)-based ACT is based on the *ex vivo* engineering and/or expansion of tumor-specific T cells followed by their reinfusion to patients. These therapies rely on the presentation of (neo)antigen-derived peptides in MHCI molecules, therefore, depend on the presence of immunogenic tumors as well as functional processing and presentation of antigenic peptides in MHC proteins. To circumvent this, newer strategies rely on the (over)expression of surface markers on cancer cells. A leading example herein is CAR-T cell therapy in which the antigen recognition capacity of antibodies is combined with T-cell activation domains, all in a single molecule (June and Sadelain, 2018).

CARs target surface-expressed antigens, enabling T-cell activation and cytotoxicity independent of MHC-peptide recognition by a TCR (Rafiq et al., 2020). In this way, immune escape through MHC-downregulation or altered MHC-peptide processing is avoided, and a single CAR construct can be used for each patient expressing the surface antigen of choice (Jacoby et al., 2019; June and Sadelain, 2018; Wirth and Kuhnel, 2017). CAR-T cell therapy is most renowned for its impressive results in hematological malignancies through targeting of CD19 and B-cell maturation antigen (BCMA), resulting in the approval of five CAR-T cell products so far: Yescarta, Kymriah, Tecartus, Breyanzi and Abecma.

The modular design of a CAR, an extracellular antigen binding part, hinge region, transmembrane part and intracellular T-cell signaling and activation domains, allows great variability. Variations for each part of the molecule are investigated to optimize CAR-T cell therapy (Fig. 3) (Benmebarek et al., 2019). While historically the main focus was on optimization of the intracellular part driving T-cell activation and proliferation, variations regarding the hinge region and the extracellular antigen-binding part have more recently proven added value in the field (Benmebarek et al., 2019; Dotti et al., 2014; Hanssens et al., 2021; June and Sadelain, 2018; Rafiq et al., 2020; Sadelain, 2017; Sadelain et al., 2013). Traditionally, CARs are designed with scFvs as the antigen-binding part (Dotti et al., 2014; Rafiq et al., 2020).

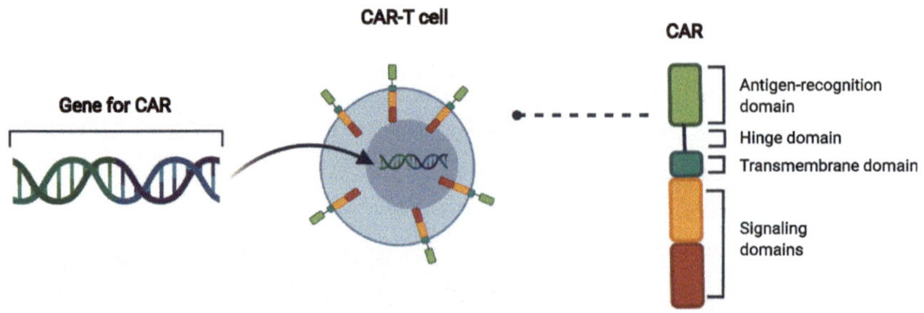

Fig. 3 Graphical representation of a CAR-T cell, the CAR-encoding gene and CAR structure. Abbreviation: CAR, chimeric antigen receptor.

Yet, some drawbacks have been associated with this design (Bannas et al., 2017; De Munter et al., 2018; Xie et al., 2019). The most important limitation of scFvs lies in their intrinsic instability due to their hydrophobic interaction regions. When not held together by disulfide bridges in the flanking constant regions in an antibody format, the organized coordination of V_H and V_L is lost, rendering them vulnerable to aggregation and domain swapping in the scFv format (Bannas et al., 2017; De Munter et al., 2018; Hanssens et al., 2021). Aggregation on the CAR-T cell surface can lead to antigen-independent T-cell activation and early exhaustion, a phenomenon known as tonic signaling (Benmebarek et al., 2019; De Munter et al., 2018; Lecocq et al., 2019a). Incorporating Nbs as the antigen-binding domain in CARs, generating nanoCARs, avoids this risk of tonic signaling, owing to the Nb's hydrophilic properties and monomeric stability (Bannas et al., 2017). Another limitation of using scFvs in CAR formats relates to their availability. Most scFvs derive from validated mAbs, as screening of entire scFv immune libraries is demanding (2020; Caruana et al., 2014; Chavez et al., 2019; Guedan et al., 2019; Sadelain et al., 2013). Also, when reformatting an antibody into an scFv-based CAR construct, linker optimization is required and adaptations may be necessary to prevent domain swapping (mispairing of variable regions) leading to CAR oligomerization (Gil and Schrum, 2013; Hanssens et al., 2021; Long et al., 2015; Xie et al., 2019). As Nbs are made of only one domain, immune libraries are relatively small to screen, allowing the evaluation of multiple candidates in a nanoCAR format (Lecocq et al., 2019a; Muyldermans, 2013). This further obviates the need for linker optimization, avoids the risk of domain swapping and allows straightforward cloning into CAR constructs, while retaining the Nb's specificity and affinity (De Munter et al., 2018; Hambach et al., 2020; Lam et al., 2020; Lecocq et al., 2019a; Xie et al., 2019). The small size of Nbs is also advantageous for production of nanoCAR encoding viral vectors. Decreasing the transgene size can enhance viral titers (Lam et al., 2020). Lastly, in terms of immunogenicity, Nbs show high sequence homology to human V_H3 family genes (Muyldermans, 2013) and at least in monomeric form were shown to be non-immunogenic (Ackaert et al., 2021). Moreover, strategies to humanize Nbs have been described (Rahbarizadeh et al., 2019; Vincke et al., 2009). While the latter is also true for murine-derived scFvs, they possess a potentially immunogenic artificial peptide-linker (Lam et al., 2020).

Multiple nanoCAR-T cell products are under preclinical investigation for diverse target antigens on both hematological malignancies as well as solid tumors (Table 5). Antigens that have been targeted with nanoCARs include CD38 (1G3) (An et al., 2018), CD19 (Wang et al., 2021), CD20

Table 5 NanoCARs.

Target	Cancer model	Compound	Specificity	Status	Reference
BCMA	MM	FHVH33 CAR-T	Human	Preclinical	Lam et al. (2020)
		LCAR-B38M (JNJ 6)		Clinical	Zhao et al. (2018)
					Xu et al. (2019)
					Madduri et al. (2020)
		PRG1801		Clinical	Han et al. (2019, 2020)
CD13	AML	Nb 157 + TIM3 CAR-T	Human	Preclinical	He et al. (2020)
CD19 and CD20	B cell lymphoma	Bispecific, CD19 + CD20 CAR-T	Human	Clinical	NCT03881761
	Burkitt lymphoma	Bispecific Nb CAR-T	Human	Preclinical	Wang et al. (2021)
CD20	Burkitt lymphoma, non-Hodgkin B cell lymphoma	CD20 CAR-T	Human	Preclinical	De Munter et al. (2020)
CD20&HER2	CD20 and HER2$^+$ Jurkat cells	Bispecific CAR-T	Human	Preclinical	De Munter et al. (2018)
CD33	AML	CD33 CAR-T	Human	Preclinical	De Munter et al. (2020)
CD38	MM, Burkitt lymphoma	1G3 CAR-T	Human	Preclinical	An et al. (2018)
		WF211, MU1067, JK36, ARTC2.2 CAR-NK			Hambach et al. (2020)
CD7	T-ALL	Mono/bivalent VHH6 CARs	Human	Preclinical	You et al. (2019)

EGFR	Epidermoid carcinoma, pharynx squamous cell carcinoma, prostatic carcinoma	7C12 Nb coupled to E5B9 epitope + anti-E5B9 uniCAR-T	Human	Preclinical	Albert et al. (2017, 2018)
	Colon, triple negative breast, non-small cell lung cancer and epidermoid carcinoma	7D12 CAR-NK mimetic		Preclinical	Gong et al. (2021)
EIIIB	Melanoma	B2 CAR-T	Murine	Preclinical	Xie et al. (2019, 2020)
GPC2	Neuroblastoma	LH7 CAR-T	Human	Preclinical	Li et al. (2017)
gp100/HLA-A2	Melanoma	GPA7 CAR-T	Human	Preclinical	Zhang et al. (2014)
HER2	Breast cancer	Oligoclonal CAR-T Nbs: RR4, RR6, RR10, RR14, RR16	Human	Preclinical	Jamnani et al. (2014)
Mucin-1	Breast cancer, squamous cell head and neck carcinoma	AR-32 CAR-T	Human	Preclinical	Bakhtiari et al. (2009) Iri-Sofla et al. (2011) Khaleghi et al. (2012)
PD-L1	B16 melanoma	A12 CAR-T	Murine	Preclinical	Xie et al. (2019, 2020)
PSMA	Prostate cancer	NBP CAR-T	Human	Preclinical	Hassani et al. (2019)
VEGFR2	293-KDR cell line	3VGR19 CAR-T	Human	Preclinical	Hajari Taheri et al. (2019)
TAG72	Colon and breast cancer	Oligoclonal CAR-T (13 Nb clones)	Human	Preclinical	Sharifzadeh et al. (2013)

Abbreviations: AML, acute myeloid leukemia; BCMA, B-cell maturation antigen; CAR, chimeric antigen receptor; EGFR, epidermal growth factor receptor; GP100, glycoprotein 100; GPC2, glypican 2; HER2, human epidermal growth factor receptor 2; HLA-A2, human leukocyte antigen A2; MM, multiple myeloma; PD-L1, programmed death ligand 1; PSMA, prostate specific membrane antigen; TAG72, tumor-associated glycoprotein 72; T-ALL, T-cell acute lymphoblastic leukemia; VEGFR2, vascular endothelial growth factor receptor 2.

(De Munter et al., 2018; Wang et al., 2021) and CD33 (De Munter et al., 2020) in hematological malignancies, and mucin-1 (AR32) (Bakhtiari et al., 2009; Iri-Sofla et al., 2011; Khaleghi et al., 2012), VEGFR2 (3VGR19) (Hajari Taheri et al., 2019) and prostate-specific membrane antigen (PSMA) (NBP) (Hassani et al., 2019) in solid tumors. Targeting the tumor microenvironment (TME) instead of TAAs might provide a more generally applicable CAR therapy. CARs targeting programmed death-ligand 1 (PD-L1) (A12) and the fibronectin splice variant EIIIB (B2) on the tumor stroma and vasculature were designed using Nbs (Xie et al., 2019). The same group later reported on an improved design, where the CAR-T cells additionally secreted Nbs or Nb fusion proteins with immunomodulatory properties against the targets PD-L1, cytotoxic T lymphocyte-associated protein 4 (CTLA4) and CD47 (Xie et al., 2020).

NanoCARs generated by combining oligoclonal Nbs against HER2 (RR4, 6,10, 14 and16) (Jamnani et al., 2014) and tumor-associated glycoprotein 72 (TAG72) (Sharifzadeh et al., 2013) were evaluated as well.

Nbs are also advantageous for the design of tandem CAR formats as no risk for domain swapping exists when multiple Nbs are expressed simultaneously from the same construct (Fig. 4). This is investigated preclinically in

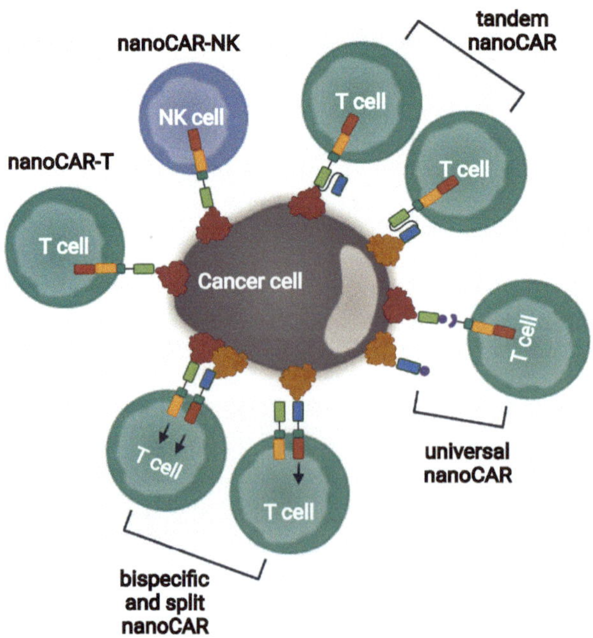

Fig. 4 Graphical representation of CAR modalities. Abbreviations: CAR, chimeric antigen receptor; NK, natural killer.

tandem CARs targeting CD20 and HER2 (De Munter et al., 2018), or targeting CD19 and CD20 (Wang et al., 2021). Bispecific and SPLIT CARs were also developed using a Nb that targets CD13 (Nb 157) and an scFv that targets T-cell immunoglobulin and mucin domain 3 (TIM3). These only become fully activated upon encounter of cells that express both targets (Fig. 4) (He et al., 2020).

CAR-T therapies depend on the availability of autologous healthy T cells. In T-cell malignancies, obtaining healthy T cells is challenging and T-cell antigens are shared between healthy and malignant cells causing fratricide of the CAR-T cells. Therefore, NK cells might serve as an alternative source for CAR therapies (Hambach et al., 2020; You et al., 2019). NanoCAR-NK cells have been developed for CD7 (VHH6) and CD38 (WF211, MU1067, JK36, ARTC2.2) (Fig. 4) (Hambach et al., 2020; You et al., 2019). Also, a non-genetic approach was explored to develop Nb-NK92MI cell conjugates as CAR-NK mimetics. The NK-cell surface was metabolically glyco-engineered allowing coupling of an anti-EGFR Nb *via* copper-free click chemistry. NK cells decorated with EGFR-targeting Nbs (7D12) showed improved antitumor efficacy in a solid tumor mouse model (Gong et al., 2021). For safety reasons, related to dose, toxicity or when encountering relapse due to antigen escape variants, universal CARs recognizing a unique peptide sequence on a separately administered targeting module rather than a TAA itself are developed (Fig. 4). Anti-EGFR (bivalent) Nbs (7C12) containing such a unique epitope are developed as the targeting modules for so-called uniCARs (Albert et al., 2017, 2018). To deal with immunogenicity concerns, nanoCARs are also being designed with human V_H domains (LH7) through screening of an engineered human antibody library (targeting glypican 2) (Li et al., 2017) or using transgenic rats encoding the complete functional human V_H gene repertoire (targeting BCMA) (FHVH33) (Lam et al., 2020). Lastly, TCR-like nanoCAR-T cells are under investigation (Zhu et al., 2020), including a Nb that targets the melanoma-related gp100/HLA-A2 complex (GPA7) (Zhang et al., 2014).

To date, nanoCAR-T products that are evaluated in clinical trials focus on hematological malignancies. Nanjing Legend Biotech/Janssen's LCAR B38M (JNJ 68284528) was awarded PRIME designation by the EMA in 2019 and is in phase III (NCT04181827). LCAR B38M encodes two Nbs in tandem, targeting different epitopes on BCMA (Bao et al., 2021; Madduri et al., 2020; Xu et al., 2019; Zhao et al., 2018). In addition, two other nanoCAR-T cell products targeting CD19/CD20 (NCT03881761) or BCMA (NCT03664661) (PRG1801) are evaluated in phase I clinical trials (Han et al., 2019, 2020).

3.2.2 Nanobody-enhanced T cell and NK cell-mediated cancer cell killing

Another type of T cell-engaging therapy, the so-called bispecific T-cell engager (BiTE) therapy, uses two targeting domains genetically linked to each other using a peptide to target surface markers expressed on cancer cells and T cells. One domain binds and activates T cells, classically through targeting of the invariant part of the TCR complex CD3 and the other domain is targeted toward a TAA of choice. When these BiTEs physically crosslink endogenous unmodified T cells with tumor cells, a pseudo-synapse is created that triggers T-cell activation. As a result, *in vitro* genetic modification, stimulation or expansion of T cells is avoided and as such, BiTE therapy represents an off-the-shelf treatment option that can be customized for multiple cancer types and their respective TAAs. To date, one BiTE is approved clinically: blinatumomab, consisting of two scFvs targeting CD19 and CD3 (Bannas et al., 2017; Einsele et al., 2020; Goebeler and Bargou, 2020).

Using scFvs in BiTE format comes with the same limitations as described earlier (among other a risk for mispairing variable regions, solubility and stability issues and potential immunogenicity) (Bannas et al., 2017). These can be circumvented by incorporating Nbs as the targeting domains, making the BiTEs more compact (Table 6) (Yang and Shah, 2020). In this context, a PEGylated BiTE consisting of an anti-CEA Nb and a conventional CD3-targeting Fab fragment was developed. PEGylation of the BiTE extended the serum half-life and delayed tumor progression more efficiently than the unPEGylated counterpart (Li et al., 2015; Pan et al., 2018). Bispecific light T-cell engagers (LiTEs) incorporating an anti-EGFR Nb and an anti-CD3 scFv domain were also developed. These LiTEs were able to activate T cells and mediate destruction of tumor cells expressing EGFR (Molgaard et al., 2018). Subsequently, LiTEs were optimized by generating an asymmetric tandem trimerbody for T-cell activation and cancer killing (ATTACK), consisting of three EGFR targeting Nbs (EGa1) genetically and alternately fused with three collagen-derived trimerization domains and coupled to an anti-CD3 scFv (OKT3). The resulting construct (EgA1 ATTACK) exhibited strong cytotoxicity when retargeting T cells to EGFR positive cells and is more efficient than the earlier reported tandem bispecific EgA1 LiTE (Harwood et al., 2017). Co-stimulation of 4-1BB shows promising antitumor potential in clinical trials, however, is often associated with toxicity related to Fc-γ receptor (FcγR) interactions. Therefore hexagonal structures, so-called trimerbodies, consisting of three EGFR- or

Table 6 Nb-based BITEs.

Target	Cancer model	Compound	Specificity	Status	Reference
CEA	Colon carcinoma	CD3 Fab + CEA Nb	Human	Preclinical	Li et al. (2015)
CEA	N/A	$1D8^{N/C}$CEA.1	Human	Preclinical	Mikkelsen et al. (2019)
EGFR	Epidermoid carcinoma	EgA1 + CD3 scFv	Human	Preclinical	Harwood et al. (2017) Molgaard et al. (2018)
EGFR	EGFR positive colon cancer	$1D8^{N/C}$EGa1	Human	Preclinical	Compte et al. (2018)
EGFR	Colon carcinoma	7D12 + 6H4	Human	Preclinical	de Bruin et al. (2016, 2017)
PD-L1	Colon cancer Patient-derived ascites	Envafolimab	Human	Preclinical	Khalique et al. (2021)

Abbreviations: CEA, carcinoembryonic antigen, EGFR, epidermal growth factor receptor; PD-L1, programmed death ligand-1.

CEA-targeting Nbs coupled to three 4-1BB-targeting scFvs have been developed. The resulting EGFR trimerbody ($1D8^{N/C}$EGa1) exhibited similar antitumor properties as IgG based 4-1BB agonists in an EGFR positive colon cancer model without the toxicity associated with IgG based 4-1BB agonists. The CEA trimerbody ($1D8^{N/C}$CEA.1) could efficiently stimulate T cells in the presence of CEA *in vitro* (Compte et al., 2018; Mikkelsen et al., 2019). Khalique (2021) developed a multimodal immunotherapy using oncolytic viruses to deliver a BiTE that targets an ICP that is highly expressed on tumor cells, while activating T cells. The BiTE contained a PD-L1 Nb and an anti-CD3 scFv and when delivered by the oncolytic virus enabled depletion of both tumor cells and tumor-promoting macrophages in patients' ascites.

A small fraction of T cells are γδT cells. Opposed to αβT cells, γδT cells are not dependent on MHC for their activation. Within the γδT cell population, Vγ9Vδ2 T cells exert potent antitumor activities. Therefore, these Vγ9Vδ2 T cells receive increasing attention in cancer therapy (Miyashita et al., 2021). de Bruin (2016) developed Nbs that bind the Vγ9Vδ2 TCR and as such activate γδT cells. An anti-EGFR Nb (7D12) was coupled to an anti-Vγ9Vδ2 TCR Nb (6H4), generating a LiTE that retained the ability to activate Vγ9Vδ2 T cells and gained the ability to potentiate killing of EGFR positive cancer cells *in vitro* and *in vivo* when co-injected with Vγ9Vδ2 T cells (de Bruin et al., 2017). Notably, these γδT cells also express the receptor CD16 (FcγRIII), a receptor that binds the Fc region of IgG1 and IgG3, and that has been studied for NK cell manipulation.

NK cells highly express CD16 and have been shown to exert ADCC when the Fc region of IgG1 mAbs that bind cancer cells, is bound by CD16 (Zahavi and Weiner, 2020). On NK cells, CD16 triggering further stimulates their proliferation and survival. This knowledge has instigated the development of Nbs that bind CD16, resulting in two Nbs that were reported to be excellent tools to generate bispecific NK-cell engagers (BiKEs) and trispecific killer engagers (TriKEs) (Table 7) (Behar et al., 2008; van Faassen et al., 2021). Both Nbs have been used to develop bispecific variants that bind cancer cells and CD16 on NK cells, achieving dual activity and increased cancer cell killing. Anti-CD16 Nbs, referred to as C21 or NRC-sdAb048, were coupled to an anti-mucin-1 Nb, an anti-CEA Nb or an anti-EGFR Nb (for C21), or an anti-EGFR Nb, an anti-HER2 Nb or anti-CD19 Nbs (for NRC-sdAb048), showing *in vitro* cytotoxicity (Dong et al., 2016; Li et al., 2018; Toffoli et al., 2021; van Faassen et al., 2021). Tumor growth delay was shown with the Nb C21-containing BiKEs that target mucin-1 and CEA in colon carcinoma xenograft mouse models, humanized by injection of peripheral blood mononuclear cells (PBMCs) (Dong et al., 2016; Li et al., 2018). A larger molecule that does not require daily administration, composed of Nb C21 coupled to the anti-CEA Nb using the CH1-Cκ heterodimerization motif, was shown to have similar tumor inhibiting activity (Rozan et al., 2013). A similar HER2-targeted construct was compared to trastuzumab, showing the ability to mediate killing of breast cancer cells with low or high HER2 expression. The first being refractory to trastuzumab therapy (Turini et al., 2014). A "knob-into-hole technology" was developed through which the anti-CD16 Nb (C21) and an anti-CEA Nb were linked to two different mutants of an IgG1 to produce a bispecific antibody chimera, showing high tumor killing potential (Li et al., 2016b). The Fabs of the HER2-targeting mAbs trastuzumab or pertuzumab have been coupled to an anti-CD16 Nb as well (Deng et al., 2018; Li et al., 2016a). The resulting constructs were compared against the mAbs showing that lower doses of the Nb fusion proteins were required to achieve equal levels of tumor cell killing *in vitro* and *in vivo*. The same construct was used to target glypican 3 positive hepatocellular carcinoma (HCC) cells, showing their effective *in vitro* lysis (Wang et al., 2018).

Table 7 Nb-based BiKEs and TriKEs.

Target	Cancer model	Compound	Specificity	Status	Reference
B7-H3	Ovarian, prostate, lung cancer cell lines and ovarian cancer xenograft model	Cam16	Human	Preclinical	Vallera et al. (2020)
CD33	Acute promyelocytic leukemia	Cam16	Human	Preclinical	Felices et al. (2020)
CEA	Ovarian and colon cancer	C21 + Nb CEA (BiSS)	Human	Preclinical	Dong et al. (2016)
CLEC12A	AML	Cam16	Human	Preclinical	Arvindam et al. (2021)
Mucin-1	Colon cancer	Nb MUC1 + Nb CEA	Human	Preclinical	Li et al. (2018)
EGFR	Epidermoid carcinoma, melanoma, colon cancer	7D12 + C21	Human	Preclinical	Toffoli et al. (2021)
EGFR HER2 CD19	Breast cancer, ovarian cancer, Raji or WIL-2S	NA1/NA2 + sdAb032, sdAb039, FMC63, sdAb040, or sdAb45	Human	Preclinical	van Faassen et al. (2021)
CEA	Colon Cancer	C17, C28 or C21 fused to bsFab (C17-C21) or C17-Fc fusion	Human	Preclinical	Rozan et al. (2013)
HER2	Ovarian cancer	Cam16	Human	Preclinical	Vallera et al. (2021)

Abbreviations: AML, acute myeloid leukemia; CEA, carcinoembryonic antigen; CLEC12A, C-type lectin domain family 12 member A; EGFR, epidermal growth factor receptor; HER2, human epidermal growth factor receptor.

Since NK cells are incapable of specific antigen recognition and cannot expand *via* activation signals alone, TriKEs have been developed. TriKEs consist of an scFv against CD16, an scFv against a TAA to enable antigen recognition by NK cells and an interleukin 15 (IL-15) moiety that stimulates NK activation and expansion. Recently, a new generation of TriKEs has been developed wherein the anti-CD16 scFv was replaced by a humanized anti-CD16 Nb (cam16). These second-generation TriKEs showed improved functionality. When using these novel TriKEs to target CD33, NK-cell proliferation, activation, tumor cell killing and tumor control in an AML tumor model were significantly improved (Felices et al., 2020). In another report, acute myeloid leukemia (AML) cells were targeted using a TriKE that binds C-type lectin domain family 12 member A. This TriKE induced *in vitro* killing of both AML cell lines and primary patient-derived AML blasts without harm to healthy cells, and reduced tumor burden in preclinical mouse models (Arvindam et al., 2021). Similarly, second-generation TriKEs have been targeted to B7 homolog 3 (B7-H3) or HER2 resulting in NK-mediated tumor cell killing in ovarian cancer xenograft models (Vallera et al., 2020, 2021).

3.3 Nanobody-mediated manipulation of immunosuppressive cell types

The TME often contains cells with immunosuppressive activity, including regulatory T cells (Tregs) and myeloid cells such as granulocytic and monocytic myeloid derived suppressor cells (MDSCs) and alternatively activated macrophages. The knowledge that these cells curtail antitumor immunity has resulted in the design of strategies to deplete or reprogram these immunosuppressive cells (Awad et al., 2018; Dees et al., 2021). Also in this context, Nbs have been used as a tool to selectively manipulate immunosuppressive cells (Table 8).

Macrophages have a dual role in the TME and are commonly referred to as TAMs. Depending on their phenotype, they exert pro- or antitumoral activities (DeNardo and Ruffell, 2019). Various Nbs have been developed to inhibit the negative effects of immunosuppressive macrophages or to polarize their phenotype toward the tumoricidal classically activated phenotype. Nbs targeting MMR were developed (Movahedi et al., 2012) and have been used for the delivery of therapeutics to deplete or repolarize tumor-promoting TAMs, as MMR is highly expressed on these cells. Repolarization of TAMs toward a tumoricidal phenotype was achieved by coupling the anti-MMR Nb to the Toll-like receptor 7/8 agonist

Table 8 Nb-based strategies to manipulate immunosuppressive cell types.

Target	Cancer model	Compound	Specificity	Status	Reference
CCR8	Colon cancer, lung cancer	Nb-I, Nb-II Fused to murine IgG2a Fc tail	Murine	Preclinical	Van Damme et al. (2021)
CD47	Melanoma	NbA4	Murine	Preclinical	Ingram et al. (2017a)
					Sockolosky et al. (2016)
	Human ovarian cancer xenograft model	HuNb1-IgG4 fusion protein	Human	Preclinical	Ma et al. (2020b)
	Human lymphoma xenograft model	CD47/CD20 bispecific antibody (BsAb)			
MMR	Tumor associated macrophages	Nb MMR fused to tSMAC	Human	Preclinical	Movahedi et al. (2012)
			Murine		De Vlaeminck et al. (2019)
NRP-1	Colon cancer	Nb1	Murine	Preclinical	De Vlaeminck et al. (2020)

Abbreviations: CCR8, chemokine receptor 8; MMR, macrophage mannose receptor; NRP-1, neuropilin-1.

imidazoquinoline (Bolli et al., 2021). To deplete MMR positive TAMs, the MMR-specific Nb was fused to an active variant of the pro-apoptotic protein second mitochondria-derived activator of caspases (SMAC). Delivery of this Nb fusion protein to macrophages or TAMs resulted in activation of caspase 3/7 followed by TAM depletion. Notably, macrophages that survived this treatment showed repolarization toward classically activated, antitumor macrophages (De Vlaeminck et al., 2019). Another Nb targeting neuropilin-1 (NRP-1), a co-receptor for semaphorins and VEGF, similarly repolarized macrophages in the TME to an antitumor phenotype without the need for an additional cargo. Nb-based blockade of NRP-1 significantly delayed tumor progression and prolonged survival in a colorectal carcinoma mouse model (De Vlaeminck et al., 2020). An important role of macrophages in the cancer immunity cycle is acquiring TAAs and presenting them to T cells.

However, APCs express a surface protein called signal regulatory protein alpha (SIRPα), a receptor that binds CD47, a 'don't eat me' signal expressed on cancer cells, thereby preventing acquisition of TAAs. Nbs (NbA4) have been used to block this CD47-SIRPα interaction to enable phagocytosis by macrophages, showing in mouse B16 melanoma that combining this Nb, vaccination and blockade of inhibitory ICPs further delayed tumor outgrowth (Ingram et al., 2017a; Sockolosky et al., 2016). Ma (2020b) developed two fusion proteins using CD47-specific Nbs. First, the CD47-specific Nbs were fused to the Fc region of IgG4, generating a CD47-specific antibody chimera (HuNb1-IgG4). Dose-dependent delay in tumor growth was observed in a human ovarian cancer xenograft model upon treatment with the Nb-Fc fusion protein. Moreover, *in vitro* assays using human red blood cells and experiments in cynomolgus monkeys showed little toxicity as opposed to anti-CD47 mAbs. Second, CD47-specific Nbs were fused with the mAb rituximab, binding hCD20, generating a bispecific construct (CD47/CD20 BsAb). The therapeutic potential of this Nb fusion protein was tested in a human lymphoma xenograft model, showing pronounced delay in tumor growth with the bispecific construct compared to rituximab monotherapy (Ma et al., 2020b). This shows that both SIRPα and CD47 are potential candidates for Nb-mediated targeted therapy.

The CC chemokine receptor 8 (CCR8) is expressed on tumor infiltrating Tregs and therefore an interesting target for the modulation of the TME. Van Damme (2021) generated two Nbs (Nb-I and Nb-II) that recognize distinct epitopes on CCR8 and formatted them to tetravalent Fc-fusions. When colorectal cancer or lung cancer bearing mice were treated with ADCC prone Nb-Fc fusion proteins, a decline in tumor infiltrating Tregs and reduced tumor growth could be observed. In contrast, ADCC deficient Nb-Fc-fusions had no effect on tumor progression suggesting that binding and blocking of CCR8 are insufficient to reduce tumor growth. When Treg depletion by ADCC competent Nb-Fc fusion proteins was combined with inhibition of programmed death-1 (PD1), an inhibitory receptor on T cells, synergistic effects were achieved.

3.4 Nanobody-mediated manipulation of cytokines in the tumor microenvironment

Cytokines play a major role in the TME and have therefore received attention as therapeutic targets (when promoting tumor growth) or agents (when mediating tumor rejection) in cancer (Berraondo et al., 2019). In this regard, Nbs have been used to neutralize cytokines or to redirect cytokines to specific cell types, reinforcing their antitumor activity (Table 9).

Table 9 Nb-based strategies to manipulate cytokines in the TME.

Target	Cancer model	Compound	Specificity	Status	Reference
CD13	Melanoma	mCD13 Nb + TNF-α Y86F or IFN-y del8	Murine	Preclinical	Huyghe et al. (2020)
CD8a	Melanoma, lung cancer	mCD8α Nb + hIL-1 βQ148G	Murine	Preclinical	Van Den Eeckhout et al. (2021)
CD20	Melanoma	mCD20 Nb + hIFNa2 Q124R hCD20 Nb + hIFNa2 Q124R	Murine Human	Preclinical	Cauwels et al. (2018a)
CEA	Colon carcinoma	CEA Nb + IL-15	Human	Preclinical	Liu et al. (2018)
Clec9A	Melanoma, breast cancer, lymphoma	Murine Clec9A + hIFNa2 Q124R Human Clec9A Nb + hIFNa2 R149A	Murine Human	Preclinical	Cauwels et al. (2018b)
G-CSF	Leukemia	VHH26	Murine Human	Preclinical	Bakherad et al. (2017)
IL-23	Not investigated	Several Nbs against subunit p19 (37D5) and p40 (22E11) of IL-23 and multivalent constructs thereof	Human	Preclinical	Desmyter et al. (2017)
TGF-β	Lung cancer cell line	Several Nbs against TGF-β were investigated	Human	Preclinical	Henry et al. (2016)
TNF-α	Breast Cancer	Anti-TNF-VHH	Human	Preclinical	Ji et al. (2017) Liu et al. (2020)
PD-L1	Pancreatic cancer	A12 + IFN-y or IL-2	Murine	Preclinical	Dougan et al. (2018)

Abbreviations: CEA, carcinoembryonic antigen; Clec9A, C-type lectin domain containing 9A; G-CSF, granulocyte-colony stimulating factor; TNF-α, tumor necrosis factor a; PD-L1, programmed death ligand-1.

Various studies have shown that TNF-α can promote tumor growth activating MDSCs and Tregs as well as stimulating therapy resistance (Benoot et al., 2021; Hu et al., 2014; Montfort et al., 2019; Zhao et al., 2012). Neutralizing anti-TNF-α Nbs showed the ability to increase response to the mitotic inhibitor Paclitaxel and the anti-VEGF mAb bevacizumab in a mouse breast cancer model (Ji et al., 2017; Liu et al., 2020). In clinically manifest tumors, transforming growth factor β (TGF-β) suppresses antitumor immune responses and increases tumor invasiveness and metastasis (Syed, 2016). A Nb that neutralizes TGF-β has been reported in 2016, yet studies on its benefit in cancer therapy have not followed (Henry et al., 2016). Granulocyte colony-stimulating factor (G-CSF) signals to neutrophils and cancer cells, facilitating tumor growth and metastasis. Blockade of G-CSF by Nb (VHH26) was beneficial in leukemia (Bakherad et al., 2017). IL-23 is overexpressed in various cancer types, has been associated with tumor growth and metastases, and suppresses antitumor functions of CTLs and NK cells (Teng et al., 2010, 2011; Yan et al., 2018). Desmyter (2017) generated Nbs against distinct epitopes of IL-23, formatted them toward multivalent Nbs and fused them to anti-human serum albumin Nbs. The multivalent anti-IL-23 Nbs could efficiently neutralize IL-23 *in vitro* and *in vivo*. However, the antitumor efficacy of these IL-23 neutralizing Nbs has not yet been reported. Recently, a trivalent IL-17 targeting Nb, sonelokimab, entered the clinic for the treatment of psoriasis (Papp et al., 2021). Since IL-17 is suggested to exert tumor-promoting proinflammatory activity (Zhao et al., 2020), it could be interesting to investigate the efficiency of this IL-17 neutralizing Nb in a cancer setting.

Cytokines with immune activating capacity are preferably delivered in a targeted manner to avoid toxicity from the cytokine therapy. Several cytokines have been delivered *via* Nbs to myeloid cells allowing them to acquire antitumor properties. Dougan (2018) targeted IFN-γ to PD-L1 on tumor-promoting TAMs, reprogramming them and inducing a delay in tumor progression. To reduce cytokine toxicity, AcTakines (activity-on-target cytokines) have been developed whereby a Nb is fused to a mutant version of a certain cytokine that exhibits low affinity for the corresponding receptor. Targeted delivery of type I IFN to the TME or DCs with respectively CD20 or Clec9A Nbs induced strong antitumor responses without toxicity, which is usually observed when using unmutated type I IFN (Cauwels et al., 2018a,b). T cells and NK cells have tumor cell killing ability and can be activated *via* cytokines. Dougan (2018) used an anti-PD-L1 Nb to deliver IL-2 to mice bearing orthotopic pancreatic tumors, showing a 50% reduction in the overall tumor burden, an outcome that was correlated with a higher ratio of CTLs over Tregs. Also, targeting of IL-1β to CTLs using CD8 targeting Nbs has been explored for use as a vaccine adjuvant. The activity of an IL-1β mutant with reduced biological activity and without systemic toxicity could be restored upon targeting using a CD8α targeting Nb (Van Den Eeckhout et al., 2020). Another strategy to increase tumor T cell levels involved targeted delivery of TNF-α to CD13 on blood vessel endothelial cells, resulting in upregulation of adhesion molecules that allowed T cells to engage and enter the tumor. Notably, co-treatment with IFN-γ targeted to CD13 ablated tumor neovasculature, in combination with T-cell approaches and led to regression of large established tumors (Huyghe et al., 2020; Van Den Eeckhout et al., 2021).

Nbs that target CEA have been fused to a human IgG1 Fc domain and linked to an IL-15 receptor α subunit (IL-15Rα)-IL-15 fusion protein. These Nb-fusions have been studied to increase the level of IL-15 in the TME as IL-15 is a potent activator of NK cells. It was shown in a xenograft colon cancer model that the resulting antitumor immune response was more potent than when IL-15 was injected systemically (Liu et al., 2018).

3.5 Nanobody-mediated immune checkpoint therapy

Inhibitory ICPs, ligand-receptor axes that dampen immune responses, are often exploited by cancer cells and their environment to paralyze effector cells and as such escape destruction by immune cells. Antibodies that interfere with the engagement of these receptors and ligands prevent immune cell exhaustion and favor tumor rejection. These molecules, also termed ICP inhibitors, became a standard of care for the treatment of various cancers and are currently the most prominent immunotherapy approach in the clinic. Following the success of conventional antibodies, Nbs have recently entered the field of ICP inhibition and a few of them even clinical trials (Table 10).

CTLA-4 (CD152) was the first ICP that has been exploited for therapy. Until now the mAb ipilimumab is used for the treatment of advanced melanoma. Several Nbs targeting CTLA-4 have been generated. Wan (2018) developed a CTLA-4 targeting Nb (Nb16) that prolonged survival in a B16 melanoma model. In a subsequent study, the same Nb was able to activate DC-HCC fusion cells to generate more potent antitumor immune responses

(Tang et al., 2019). Another anti-CTLA-4 Nb (H11) developed by Ingram (2018) showed superior CTLA-4 blocking efficiency compared to CTLA-4 mAbs, however had negligible effect on tumor progression. When coupled to a murine IgG2a Fc region, generating an antibody chimera, similar effects to full size CTLA-4 mAbs could be achieved, suggesting that Fc-mediated effector functions are indispensable to induce substantial antitumor immune responses when targeting CTLA-4. In 2020, a first anti-CTLA-4 Nb comprising an Fc region (KN044) entered clinical testing (NCT04126590).

PD1 and its ligand PD-L1 (B7-H1, CD274) gained much attention since blockade of this ICP axis leads to robust antitumor immune responses with an improved safety profile (Naidoo et al., 2015). Several Nbs that target either PD1 or PD-L1 have been developed. We have developed an anti-human PD-L1 Nb (K2) that facilitates tumor cell killing and T-cell activation by DCs in cases where mAbs fail (Broos et al., 2019a,b). By increasing the valency of this Nb, we improved the blocking efficiency equally to the clinically approved mAb avelumab that competes with the Nb for binding to PD-L1. Moreover, we showed that generating an antibody chimera using an IgG1 Fc region enhanced immune cell activation and tumor cell killing (Awad et al., 2021). Similarly, Zhang (2017) developed an anti-PD-L1 Nb that restores T-cell activation and demonstrated that an antibody chimera thereof (KN035) was able to delay tumor growth *in vivo*. In 2020, this Nb entered clinical testing under the name envafolimab, showing therapeutic efficacy upon subcutaneous administration in patients with solid tumors (Li et al., 2021; Papadopoulos et al., 2021). Various Nbs (Nb97, Nb-B20, Nb11) that bind to PD1 and block the interaction with PD-L1 have been described (Silva-Pilipich et al., 2020; Wen et al., 2020; Xian et al., 2019).

Various other ICPs have been described and contribute to immune evasion and cancer progression. Of these, lymphocyte activation gene 3 (LAG3, CD223) and TIM3 (CD366) were identified as interesting targets for ICP therapy (Acharya et al., 2020; Lecocq et al., 2020). We developed Nbs that are able to bind mouse or human LAG3, yet their therapy efficacy remains to be determined (Lecocq et al., 2019b, 2021a,b). Homayouni (2016) developed anti-TIM3 Nbs and showed that they inhibit the proliferation of an AML cell line.

Systemic ICP therapy with mAbs often leads to unsatisfying response rates and a multitude of toxicities. Therefore, efforts have been made to

Table 10 Nb-based ICP therapy.

Target	Cancer model	Compound	Specificity	Status	Reference
CTLA-4	Melanoma	H11	Murine	Preclinical	Ingram et al. (2018) Xie et al. (2020)
CTLA-4	Melanoma, breast cancer, hepatocarcinoma	Nb16	Murine	Preclinical	Wan et al. (2018)
			Human		Tang et al. (2019)
CTLA-4	Colon cancer, lymphoma	5E03	Murine	Preclinical	Gurbatri et al. (2020)
CTLA-4	Locally advanced or metastatic solid tumors	KN044	Human	Clinical	NCT04126590
PD-L1	Melanoma, breast cancer	K2	Human	Preclinical	Broos et al. (2019a) Awad et al. (2021)
PD-L1	Melanoma, solid tumors	KN035 (Envafolimab)	Human	Clinical	Zhang et al. (2017) Papadopoulos et al. (2021) Li et al. (2021)
PD-L1	Colon cancer, lymphoma	5DXW	Murine	Preclinical	Gurbatri et al. (2020)
PD-L1	Colon cancer, melanoma	A12	Murine	Preclinical	Ingram et al. (2017b) Xie et al. (2020)
PD-L1	Colon cancer	Tetravalent PD-L1 Nb-Fc	Human	Preclinical	Ma et al. (2020a)
PD-1	PD1 reporter assay	MY2938 consists of two Nb97 fused to human serum albumin	Human	Preclinical	Xian et al. (2019)
PD-1	PD-L1 positive pancreas cancer cells	Nb-B20	Human	Preclinical	Wen et al. (2020)

Continued

Table 10 Nb-based ICP therapy.—cont'd

Target	Cancer model	Compound	Specificity	Status	Reference
PD-1	Colon cancer	Nb11	Murine Human	Preclinical	Silva-Pilipich et al. (2020)
TIM3	TIM3 positive monocytic leukemia cells	Nb60	Human	Preclinical	Homayouni et al. (2016)
TIGIT	PBMC assay	Tetravalent TIGIT NbA-NbB-Fc	Human	Preclinical	Ma et al. (2020a)
CD47	Melanoma, colon cancer	A4	Murine	Preclinical	Xie et al. (2020)
PD-L1/ CTLA-4	Advanced solid tumors, thymic cancer	KN046	Human	Clinical	Coward et al. (2019) NCT03529526 NCT04469725
PD-L1/ TIGIT	PBMC assay	PD-L1/TIGIT BsAb	Human	Preclinical	Ma et al. (2020a)
4-1BB/ PD-L1	Colon cancer	PM1003	Human	Preclinical	Zhai et al. (2021)

Abbreviations: CTLA-4, cytotoxic T-lymphocyte-associated antigen 4; PBMC, peripheral blood mononuclear cells; PD-1, programmed cell death protein-1; PD-L1, programmed death ligand-1; TIM3, T cell immunoglobulin and mucin domain-containing protein 3; TIGIT, T cell immunoglobulin and ITIM domain.

enhance ICP therapy and to restrict the drugs to the tumor site. Various reports addressed combination therapies and novel delivery approaches of Nbs for ICP therapy. Due to their small size, Nbs can be easily encoded in various expression vectors. Since no complex posttranslational modifications are required for Nb synthesis and function, prokaryotic expression systems can be used for Nb delivery. Bacteria have been transformed with plasmids encoding anti-CTLA-4 (5E03) and anti-PD-L1 (5DXW) Nbs. These were administered to tumors, resulting in local Nb production, improved anti-tumor activity and even abscopal effects (Gurbatri et al., 2020).

Also, CAR-T cells have been modified to secrete a Nb-Fc fusion targeting CTLA-4 (H11) or a Nb targeting PD-L1 (A12), which resulted in elevated CAR T-cell levels in spleen, lymph nodes and B16 tumors but did not significantly impact survival. When PD-L1 nanoCAR-T cells secreted a Nb-Fc fusion targeting CD47 (A4), tumor growth was significantly reduced (Xie et al., 2020).

Several viral approaches for the delivery of ICP blocking Nbs have been evaluated as well. Anti-PD1 Nbs (Nb11) were encoded into adeno-associated viruses and administered intravenously into mice, resulting in long term Nb expression without evidence of toxicity. When these mice were challenged with colorectal tumor cells, survival was significantly improved (Silva-Pilipich et al., 2020). We encoded anti-human PD-L1 Nbs (K2) and variants thereof in LVs. These were used to transduce melanoma cells and enabled tumor cell killing when these melanoma cells were co-cultured with PBMCs (Awad et al., 2021).

Expression of ICPs is dynamic and targeting one ICP can lead to the compensatory upregulation of others (Huang et al., 2017; Lecocq et al., 2021a). Several Nbs have been developed either targeting two inhibitory ICPs or comprising an immune stimulatory molecule. Ma et al. (2020a) generated tetravalent PD-L1 and T cell immunoglobulin and ITIM domain (TIGIT) Nbs and showed that these Nbs were able to block respectively the PD1/PD-L1 or TIGIT/CD155 interactions. These Nb constructs were used to create a bispecific antibody that retained blocking efficiency and exhibited synergistic effects *in vitro*. Anti-PD-L1 (C-Ye-18) and anti-4-1BB (L-Yr-16) Nbs were fused *via* a silent Fc region, resulting in PM1003, a bispecific Nb construct, able to suppress colon cancer growth with little hepatotoxicity (Zhai et al., 2021). To date, an antibody chimera containing an IgG1 Fc region coupled to anti-PD-L1 and anti-CTLA-4 Nbs (KN046) entered clinical testing. The bispecific Nb-Fc fusion showed an acceptable safety profile and promising efficacy in a phase I trial (NCT03529526) and is further investigated in a phase II trial (NCT04469725) (Coward et al., 2019).

4. Conclusion

The use of Nbs has high potential to improve surgery, chemotherapy, radiation therapy and immunotherapy. The Nbs' unique structure and the resulting fast clearance make them viable as tracers for surgery and targeted delivery of toxins or radionuclides to the tumor site. For immunotherapeutic approaches, Nbs are currently explored for cancer vaccination, direct and indirect manipulation of cancer killing T cells and NK cells as well as modulation of the immunosuppressive TME. Two approaches, *i.e.*, the use of Nbs in CAR development and ICP therapy have rapidly evolved, resulting in various ongoing clinical trials. These trials together with preclinical research on Nbs becoming more and more vivid show that Nbs alike mAbs have earned credibility as a high potential biological drug. The possibility to stratify patients before and during therapy based on same-day noninvasive imaging of the envisaged target antigen with the Nb used for therapy further adds to the Nbs' appeal. We strongly believe that in the future, Nb-based imaging to map the TME combined with subsequent personalized Nb-based cancer therapy will become a routine workflow in oncology.

References

Anon., 2020. FDA okays second CAR-T for Kite. Nat. Biotechnol. 38, 1012.

Abulrob, A., Sprong, H., Van Bergen en Henegouwen, P., Stanimirovic, D., 2005. The blood-brain barrier transmigrating single domain antibody: mechanisms of transport and antigenic epitopes in human brain endothelial cells. J. Neurochem. 95, 1201–1214.

Acharya, N., Sabatos-Peyton, C., Anderson, A.C., 2020. Tim-3 finds its place in the cancer immunotherapy landscape. J. Immunother. Cancer 8.

Ackaert, C., Smiejkowska, N., Xavier, C., Sterckx, Y.G.J., Denies, S., Stijlemans, B., Elkrim, Y., Devoogdt, N., Caveliers, V., Lahoutte, T., Muyldermans, S., Breckpot, K., Keyaerts, M., 2021. Immunogenicity risk profile of nanobodies. Front. Immunol. 12, 632687.

Ageichik, A., Buchholz, C.J., Collins, M.K., 2011. Lentiviral vectors targeted to MHC II are effective in immunization. Hum. Gene Ther. 22, 1249–1254.

Albert, S., Arndt, C., Feldmann, A., Bergmann, R., Bachmann, D., Koristka, S., Ludwig, F., Ziller-Walter, P., Kegler, A., Gartner, S., Schmitz, M., Ehninger, A., Cartellieri, M., Ehninger, G., Pietzsch, H.J., Pietzsch, J., Steinbach, J., Bachmann, M., 2017. A novel nanobody-based target module for retargeting of T lymphocytes to EGFR-expressing cancer cells via the modular UniCAR platform. Oncoimmunology 6, e1287246.

Albert, S., Arndt, C., Koristka, S., Berndt, N., Bergmann, R., Feldmann, A., Schmitz, M., Pietzsch, J., Steinbach, J., Bachmann, M., 2018. From mono- to bivalent: improving theranostic properties of target modules for redirection of UniCAR T cells against EGFR-expressing tumor cells in vitro and in vivo. Oncotarget 9, 25597–25616.

Altintas, I., Heukers, R., van der Meel, R., Lacombe, M., Amidi, M., van Bergen En Henegouwen, P.M., Hennink, W.E., Schiffelers, R.M., Kok, R.J., 2013. Nanobody-albumin nanoparticles (NANAPs) for the delivery of a multikinase inhibitor 17864 to EGFR overexpressing tumor cells. J. Control. Release 165, 110–118.

An, N., Hou, Y.N., Zhang, Q.X., Li, T., Zhang, Q.L., Fang, C., Chen, H., Lee, H.C., Zhao, Y.J., Du, X., 2018. Anti-multiple myeloma activity of nanobody-based anti-CD38 chimeric antigen receptor T cells. Mol. Pharm. 15, 4577–4588.

Arvindam, U.S., van Hauten, P.M.M., Schirm, D., Schaap, N., Hobo, W., Blazar, B.R., Vallera, D.A., Dolstra, H., Felices, M., Miller, J.S., 2021. A trispecific killer engager molecule against CLEC12A effectively induces NK-cell mediated killing of AML cells. Leukemia 35, 1586–1596.

Awad, R.M., De Vlaeminck, Y., Maebe, J., Goyvaerts, C., Breckpot, K., 2018. Turn back the TIMe: targeting tumor infiltrating myeloid cells to revert cancer progression. Front. Immunol. 9, 1977.

Awad, R.M., Lecocq, Q., Zeven, K., Ertveldt, T., De Beck, L., Ceuppens, H., Broos, K., De Vlaeminck, Y., Goyvaerts, C., Verdonck, M., Raes, G., Van Parys, A., Cauwels, A., Keyaerts, M., Devoogdt, N., Breckpot, K., 2021. Formatting and gene-based delivery of a human PD-L1 single domain antibody for immune checkpoint blockade. Mol. Ther. Methods Clin. Dev. 22, 172–182.

Bakherad, H., Gargari, S.L.M., Sepehrizadeh, Z., Aghamollaei, H., Taheri, R.A., Torshabi, M., Yazdi, M.T., Ebrahimizadeh, W., Setayesh, N., 2017. Identification and in vitro characterization of novel nanobodies against human granulocyte colony-stimulating factor receptor to provide inhibition of G-CSF function. Biomed. Pharmacother. 93, 245–254.

Bakhtiari, S.H., Rahbarizadeh, F., Hasannia, S., Ahmadvand, D., Iri-Sofla, F.J., Rasaee, M.J., 2009. Anti-MUC1 nanobody can redirect T-body cytotoxic effector function. Hybridoma (Larchmt) 28, 85–92.

Banihashemi, S.R., Rahbarizadeh, F., Zavaran Hosseini, A., Ahmadvand, D., Khoshtinat Nikkhoi, S., 2021. Liposome-based nanocarriers loaded with anthrax lethal factor and armed with anti-CD19 VHH for effectively inhibiting MAPK pathway in B cells. Int. Immunopharmacol. 100, 107927.

Bannas, P., Hambach, J., Koch-Nolte, F., 2017. Nanobodies and nanobody-based human heavy chain antibodies as antitumor therapeutics. Front. Immunol. 8, 1603.

Bao, C., Gao, Q., Li, L.L., Han, L., Zhang, B., Ding, Y., Song, Z., Zhang, R., Zhang, J., Wu, X.H., 2021. The application of nanobody in CAR-T therapy. Biomolecules 11.

Behar, G., Siberil, S., Groulet, A., Chames, P., Pugniere, M., Boix, C., Sautes-Fridman, C., Teillaud, J.L., Baty, D., 2008. Isolation and characterization of anti-FcgammaRIII (CD16) llama single-domain antibodies that activate natural killer cells. Protein Eng. Des. Sel. 21, 1–10.

Behdani, M., Zeinali, S., Karimipour, M., Khanahmad, H., Schoonooghe, S., Aslemarz, A., Seyed, N., Moazami-Godarzi, R., Baniahmad, F., Habibi-Anbouhi, M., Hassanzadeh-Ghassabeh, G., Muyldermans, S., 2013. Development of VEGFR2-specific nanobody Pseudomonas exotoxin A conjugated to provide efficient inhibition of tumor cell growth. N. Biotechnol. 30, 205–209.

Benmebarek, M.R., Karches, C.H., Cadilha, B.L., Lesch, S., Endres, S., Kobold, S., 2019. Killing mechanisms of chimeric antigen receptor (CAR) T cells. Int. J. Mol. Sci. 20.

Benoot, T., Piccioni, E., De Ridder, K., Goyvaerts, C., 2021. TNFalpha and immune checkpoint inhibition: friend or foe for lung cancer? Int. J. Mol. Sci. 22.

Berraondo, P., Sanmamed, M.F., Ochoa, M.C., Etxeberria, I., Aznar, M.A., Perez-Gracia, J.L., Rodriguez-Ruiz, M.E., Ponz-Sarvise, M., Castanon, E., Melero, I., 2019. Cytokines in clinical cancer immunotherapy. Br. J. Cancer 120, 6–15.

Bolli, E., D'Huyvetter, M., Murgaski, A., Berus, D., Stange, G., Clappaert, E.J., Arnouk, S., Pombo Antunes, A.R., Krasniqi, A., Lahoutte, T., Goncalves, A., Vuylsteke, M., Raes, G., Devoogdt, N., Movahedi, K., Van Ginderachter, J.A., 2019. Stromal-targeting radioimmunotherapy mitigates the progression of therapy-resistant tumors. J. Control. Release 314, 1–11.
Bolli, E., Scherger, M., Arnouk, S.M., Pombo Antunes, A.R., Straßburger, D., Urschbach, M., Stickdorn, J., De Vlaminck, K., Movahedi, K., Räder, H.J., Hernot, S., Besenius, P., Van Ginderachter, J.A., Nuhn, L., 2021. Targeted repolarization of tumor-associated macrophages via imidazoquinoline-linked nanobodies. Adv. Sci. 8, 2004574.
Bonehill, A., Heirman, C., Thielemans, K., 2005. Genetic approaches for the induction of a CD4+ T cell response in cancer immunotherapy. J. Gene Med. 7, 686–695.
Bonifaz, L.C., Bonnyay, D.P., Charalambous, A., Darguste, D.I., Fujii, S., Soares, H., Brimnes, M.K., Moltedo, B., Moran, T.M., Steinman, R.M., 2004. In vivo targeting of antigens to maturing dendritic cells via the DEC-205 receptor improves T cell vaccination. J. Exp. Med. 199, 815–824.
Botelho, N.K., Tschumi, B.O., Hubbell, J.A., Swartz, M.A., Donda, A., Romero, P., 2019. Combination of synthetic long peptides and XCL1 fusion proteins results in superior tumor control. Front. Immunol. 10, 294.
Broos, K., Lecocq, Q., Keersmaecker, B., Raes, G., Corthals, J., Lion, E., Thielemans, K., Devoogdt, N., Keyaerts, M., Breckpot, K., 2019a. Single domain antibody-mediated blockade of programmed death-ligand 1 on dendritic cells enhances CD8 T-cell activation and cytokine production. Vaccines (Basel) 7, 85.
Broos, K., Lecocq, Q., Xavier, C., Bridoux, J., Nguyen, T.T., Corthals, J., Schoonooghe, S., Lion, E., Raes, G., Keyaerts, M., Devoogdt, N., Breckpot, K., 2019b. Evaluating a single domain antibody targeting human PD-L1 as a nuclear imaging and therapeutic agent. Cancers (Basel) 11, 872.
Caruana, I., Diaconu, I., Dotti, G., 2014. From monoclonal antibodies to chimeric antigen receptors for the treatment of human malignancies. Semin. Oncol. 41, 661–666.
Cauwels, A., Van Lint, S., Garcin, G., Bultinck, J., Paul, F., Gerlo, S., Van der Heyden, J., Bordat, Y., Catteeuw, D., De Cauwer, L., Rogge, E., Verhee, A., Uze, G., Tavernier, J., 2018a. A safe and highly efficient tumor-targeted type I interferon immunotherapy depends on the tumor microenvironment. Oncoimmunology 7, e1398876.
Cauwels, A., Van Lint, S., Paul, F., Garcin, G., De Koker, S., Van Parys, A., Wueest, T., Gerlo, S., Van der Heyden, J., Bordat, Y., Catteeuw, D., Rogge, E., Verhee, A., Vandekerckhove, B., Kley, N., Uze, G., Tavernier, J., 2018b. Delivering type I interferon to dendritic cells empowers tumor eradication and immune combination treatments. Cancer Res. 78, 463–474.
Chavez, J.C., Bachmeier, C., Kharfan-Dabaja, M.A., 2019. CAR T-cell therapy for B-cell lymphomas: clinical trial results of available products. Ther. Adv. Hematol. 10, 2040620719841581.
Cire, S., Da Rocha, S., Yao, R., Fisson, S., Buchholz, C.J., Collins, M.K., Galy, A., 2014. Immunization of mice with lentiviral vectors targeted to MHC class II+ cells is due to preferential transduction of dendritic cells in vivo. PLoS One 9, e101644.
Compte, M., Harwood, S.L., Munoz, I.G., Navarro, R., Zonca, M., Perez-Chacon, G., Erce-Llamazares, A., Merino, N., Tapia-Galisteo, A., Cuesta, A.M., Mikkelsen, K., Caleiras, E., Nunez-Prado, N., Aznar, M.A., Lykkemark, S., Martinez-Torrecuadrada, J., Melero, I., Blanco, F.J., Bernardino de la Serna, J., Zapata, J.M., Sanz, L., Alvarez-Vallina, L., 2018. A tumor-targeted trimeric 4-1BB-agonistic antibody induces potent anti-tumor immunity without systemic toxicity. Nat. Commun. 9, 4809.
Cortez-Retamozo, V., Backmann, N., Senter, P.D., Wernery, U., De Baetselier, P., Muyldermans, S., Revets, H., 2004. Efficient cancer therapy with a nanobody-based conjugate. Cancer Res. 64, 2853–2857.

Coulie, P.G., Van den Eynde, B.J., van der Bruggen, P., Boon, T., 2014. Tumour antigens recognized by T lymphocytes: at the core of cancer immunotherapy. Nat. Rev. Cancer 14, 135–146.

Coward, J., Ganju, V., Behzadigohar, R., Kwong, K., Xu, J., Van, H., Kong, P., Yang, F., Chen, L., Guo, K., Liu, M., Zhu, D., Donato, L.K., Xu, T., Richardson, G.E., 2019. Preliminary safety, efficacy, and pharmacokinetics (PK) results of KN046 (bispecific anti-PD-L1/CTLA4) from a first-in-human study in subjects with advanced solid tumors. J. Clin. Oncol. 37, 2554.

Crowley, S.J., Bruck, P.T., Bhuiyan, M.A., Mitchell-Gears, A., Walsh, M.J., Zhangxu, K., Ali, L.R., Jeong, H.J., Ingram, J.R., Knipe, D.M., Ploegh, H.L., Dougan, M., Dougan, S.K., 2020. Neoleukin-2 enhances anti-tumour immunity downstream of peptide vaccination targeted by an anti-MHC class II VHH. Open Biol. 10, 190235.

Cruz, L.J., Rosalia, R.A., Kleinovink, J.W., Rueda, F., Lowik, C.W., Ossendorp, F., 2014. Targeting nanoparticles to CD40, DEC-205 or CD11c molecules on dendritic cells for efficient CD8(+) T cell response: a comparative study. J. Control. Release 192, 209–218.

Czernin, J., Current, K., Mona, C.E., Nyiranshuti, L., Hikmat, F., Radu, C.G., Luckerath, K., 2021. Immune-checkpoint blockade enhances (225)Ac-PSMA617 efficacy in a mouse model of prostate cancer. J. Nucl. Med. 62, 228–231.

Debie, P., Hernot, S., 2019. Emerging fluorescent molecular tracers to guide intra-operative surgical decision-making. Front. Pharmacol. 10, 510.

Debie, P., Van Quathem, J., Hansen, I., Bala, G., Massa, S., Devoogdt, N., Xavier, C., Hernot, S., 2017. Effect of dye and conjugation chemistry on the biodistribution profile of near-infrared-labeled nanobodies as tracers for image-guided surgery. Mol. Pharm. 14, 1145–1153.

Debie, P., Vanhoeij, M., Poortmans, N., Puttemans, J., Gillis, K., Devoogdt, N., Lahoutte, T., Hernot, S., 2018. Improved debulking of peritoneal tumor implants by near-infrared fluorescent nanobody image guidance in an experimental mouse model. Mol. Imaging Biol. 20, 361–367.

Debie, P., Lafont, C., Defrise, M., Hansen, I., van Willigen, D.M., van Leeuwen, F.W.B., Gijsbers, R., D'Huyvetter, M., Devoogdt, N., Lahoutte, T., Mollard, P., Hernot, S., 2020. Size and affinity kinetics of nanobodies influence targeting and penetration of solid tumours. J. Control. Release 317, 34–42.

Debie, P., Declerck, N.B., van Willigen, D., Huygen, C.M., De Sloovere, B., Mateusiak, L., Bridoux, J., Puttemans, J., Devoogdt, N., van Leeuwen, F.W.B., Hernot, S., 2021. The design and preclinical evaluation of a single-label bimodal nanobody tracer for image-guided surgery. Biomolecules 11, 360.

de Bruijn, H.S., Mashayekhi, V., Schreurs, T.J.L., van Driel, P., Strijkers, G.J., van Diest, P.J., Lowik, C., Seynhaeve, A.L.B., Hagen, T., Prompers, J.J., Henegouwen, P., Robinson, D.J., Oliveira, S., 2020. Acute cellular and vascular responses to photodynamic therapy using EGFR-targeted nanobody-photosensitizer conjugates studied with intravital optical imaging and magnetic resonance imaging. Theranostics 10, 2436–2452.

de Bruin, R.C.G., Lougheed, S.M., van der Kruk, L., Stam, A.G., Hooijberg, E., Roovers, R.C., van Bergen En Henegouwen, P.M.P., Verheul, H.M.W., de Gruijl, T.D., van der Vliet, H.J., 2016. Highly specific and potently activating Vgamma9Vdelta2-T cell specific nanobodies for diagnostic and therapeutic applications. Clin. Immunol. 169, 128–138.

de Bruin, R.C.G., Veluchamy, J.P., Lougheed, S.M., Schneiders, F.L., Lopez-Lastra, S., Lameris, R., Stam, A.G., Sebestyen, Z., Kuball, J., Molthoff, C.F.M., Hooijberg, E., Roovers, R.C., Santo, J.P.D., van Bergen En Henegouwen, P.M.P., Verheul, H.M.W., de Gruijl, T.D., van der Vliet, H.J., 2017. A bispecific nanobody approach to leverage the potent and widely applicable tumor cytolytic capacity of Vgamma9Vdelta2-T cells. Oncoimmunology 7, e1375641.

Dees, S., Ganesan, R., Singh, S., Grewal, I.S., 2021. Regulatory T cell targeting in cancer: emerging strategies in immunotherapy. Eur. J. Immunol. 51, 280–291.

De Groeve, K., Deschacht, N., De Koninck, C., Caveliers, V., Lahoutte, T., Devoogdt, N., Muyldermans, S., De Baetselier, P., Raes, G., 2010. Nanobodies as tools for in vivo imaging of specific immune cell types. J. Nucl. Med. 51, 782–789.

De Groof, T.W.M., Mashayekhi, V., Fan, T.S., Bergkamp, N.D., Sastre Toraño, J., van Senten, J.R., Heukers, R., Smit, M.J., Oliveira, S., 2019. Nanobody-targeted photodynamic therapy selectively kills viral GPCR-expressing glioblastoma cells. Mol. Pharm. 16, 3145–3156.

Dekempeneer, Y., Back, T., Aneheim, E., Jensen, H., Puttemans, J., Xavier, C., Keyaerts, M., Palm, S., Albertsson, P., Lahoutte, T., Caveliers, V., Lindegren, S., D'Huyvetter, M., 2019. Labeling of anti-HER2 nanobodies with astatine-211: optimization and the effect of different coupling reagents on their in vivo behavior. Mol. Pharm. 16, 3524–3533.

Dekempeneer, Y., Caveliers, V., Ooms, M., Maertens, D., Gysemans, M., Lahoutte, T., Xavier, C., Lecocq, Q., Maes, K., Covens, P., Miller, B.W., Bruchertseifer, F., Morgenstern, A., Cardinaels, T., D'Huyvetter, M., 2020. Therapeutic efficacy of (213)Bi-labeled sdAbs in a preclinical model of ovarian cancer. Mol. Pharm. 17, 3553–3566.

Deken, M.M., Kijanka, M.M., Beltrán Hernández, I., Slooter, M.D., de Bruijn, H.S., van Diest, P.J., van Bergen En Henegouwen, P.M.P., Lowik, C.W.G.M., Robinson, D.J., Vahrmeijer, A.L., Oliveira, S., 2020. Nanobody-targeted photodynamic therapy induces significant tumor regression of trastuzumab-resistant HER2-positive breast cancer, after a single treatment session. J. Control. Release 323, 269–281.

De Munter, S., Ingels, J., Goetgeluk, G., Bonte, S., Pille, M., Weening, K., Kerre, T., Abken, H., Vandekerckhove, B., 2018. Nanobody based dual specific CARs. Int. J. Mol. Sci. 19, 403.

De Munter, S., Van Parys, A., Bral, L., Ingels, J., Goetgeluk, G., Bonte, S., Pille, M., Billiet, L., Weening, K., Verhee, A., Van der Heyden, J., Taghon, T., Leclercq, G., Kerre, T., Tavernier, J., Vandekerckhove, B., 2020. Rapid and effective generation of nanobody based CARs using PCR and Gibson assembly. Int. J. Mol. Sci. 21, 883.

DeNardo, D.G., Ruffell, B., 2019. Macrophages as regulators of tumour immunity and immunotherapy. Nat. Rev. Immunol. 19, 369–382.

Deng, W., Liu, J., Pan, H., Li, L., Zhou, C., Wang, X., Shu, R., Dong, B., Cao, D., Li, Q., Wang, Z., 2018. A bispecific antibody based on pertuzumab Fab has potent antitumor activity. J. Immunother. 41, 1–8.

Desmyter, A., Spinelli, S., Boutton, C., Saunders, M., Blachetot, C., de Haard, H., Denecker, G., Van Roy, M., Cambillau, C., Rommelaere, H., 2017. Neutralization of human interleukin 23 by multivalent nanobodies explained by the structure of cytokine-nanobody complex. Front. Immunol. 8, 884.

De Veirman, K., Puttemans, J., Krasniqi, A., Ertveldt, T., Hanssens, H., Romao, E., Hose, D., Goyvaert, C., Vlummens, P., Muyldermans, S., Breckpot, K., Bruchertseifer, F., Morgenstern, A., D'Huyvetter, M., Devoogdt, N., 2021. CS1-specific single-domain antibodies labeled with Actinium-225 prolong survival and increase CD8+ T cells and PD-L1 expression in Multiple Myeloma. Oncoimmunology 10, 2000699.

De Vlaeminck, Y., Lecocq, Q., Giron, P., Heirman, C., Geeraerts, X., Bolli, E., Movahedi, K., Massa, S., Schoonooghe, S., Thielemans, K., Goyvaerts, C., Van Ginderachter, J.A., Breckpot, K., 2019. Single-domain antibody fusion proteins can target and shuttle functional proteins into macrophage mannose receptor expressing macrophages. J. Control. Release 299, 107–120.

De Vlaeminck, Y., Bonelli, S., Awad, R.M., Dewilde, M., Rizzolio, S., Lecocq, Q., Bolli, E., Santos, A.R., Laoui, D., Schoonooghe, S., Tamagnone, L., Goyvaerts, C., Mazzone, M., Breckpot, K., Van Ginderachter, J.A., 2020. Targeting neuropilin-1 with nanobodies reduces colorectal carcinoma development. Cancers (Basel) 12, 3582.

Dhodapkar, M.V., Sznol, M., Zhao, B., Wang, D., Carvajal, R.D., Keohan, M.L., Chuang, E., Sanborn, R.E., Lutzky, J., Powderly, J., Kluger, H., Tejwani, S., Green, J., Ramakrishna, V., Crocker, A., Vitale, L., Yellin, M., Davis, T., Keler, T., 2014. Induction of antigen-specific immunity with a vaccine targeting NY-ESO-1 to the dendritic cell receptor DEC-205. Sci. Transl. Med. 6, 232ra251.

D'Huyvetter, M., Vincke, C., Xavier, C., Aerts, A., Impens, N., Baatout, S., De Raeve, H., Muyldermans, S., Caveliers, V., Devoogdt, N., Lahoutte, T., 2014. Targeted radionuclide therapy with A 177Lu-labeled anti-HER2 nanobody. Theranostics 4, 708–720.

D'Huyvetter, M., De Vos, J., Xavier, C., Pruszynski, M., Sterckx, Y.G.J., Massa, S., Raes, G., Caveliers, V., Zalutsky, M.R., Lahoutte, T., Devoogdt, N., 2017. (131) I-labeled anti-HER2 camelid sdAb as a theranostic tool in cancer treatment. Clin. Cancer Res. 23, 6616–6628.

D'Huyvetter, M., Vos, J., Caveliers, V., Vaneycken, I., Heemskerk, J., Duhoux, F.P., Fontaine, C., Vanhoeij, M., Windhorst, A.D., Aa, F.V., Hendrikse, N.H., Eersels, J.L.E., Everaert, H., Gykiere, P., Devoogdt, N., Raes, G., Lahoutte, T., Keyaerts, M., 2021. Phase I trial of (131)I-GMIB-anti-HER2-VHH1, a new promising candidate for HER2-targeted radionuclide therapy in breast cancer patients. J. Nucl. Med. 62, 1097–1105.

Dong, B., Zhou, C., He, P., Li, J., Chen, S., Miao, J., Li, Q., Wang, Z., 2016. A novel bispecific antibody, BiSS, with potent anti-cancer activities. Cancer Biol. Ther. 17, 364–370.

Dotti, G., Gottschalk, S., Savoldo, B., Brenner, M.K., 2014. Design and development of therapies using chimeric antigen receptor-expressing T cells. Immunol. Rev. 257, 107–126.

Dougan, M., Ingram, J.R., Jeong, H.J., Mosaheb, M.M., Bruck, P.T., Ali, L., Pishesha, N., Blomberg, O., Tyler, P.M., Servos, M.M., Rashidian, M., Nguyen, Q.D., von Andrian, U.H., Ploegh, H.L., Dougan, S.K., 2018. Targeting cytokine therapy to the pancreatic tumor microenvironment using PD-L1-specific VHHs. Cancer Immunol. Res. 6, 389–401.

Duray, E., Lejeune, M., Baron, F., Beguin, Y., Devoogdt, N., Krasniqi, A., Lauwers, Y., Zhao, Y.J., D'Huyvetter, M., Dumoulin, M., Caers, J., 2021. A non-internalised CD38-binding radiolabelled single-domain antibody fragment to monitor and treat multiple myeloma. J. Hematol. Oncol. 14, 183.

Einsele, H., Borghaei, H., Orlowski, R.Z., Subklewe, M., Roboz, G.J., Zugmaier, G., Kufer, P., Iskander, K., Kantarjian, H.M., 2020. The BiTE (bispecific T-cell engager) platform: development and future potential of a targeted immuno-oncology therapy across tumor types. Cancer 126, 3192–3201.

Elzoghby, A.O., Samy, W.M., Elgindy, N.A., 2012. Albumin-based nanoparticles as potential controlled release drug delivery systems. J. Control. Release 157, 168–182.

Epenetos, A.A., Snook, D., Durbin, H., Johnson, P.M., Taylor-Papadimitriou, J., 1986. Limitations of radiolabeled monoclonal antibodies for localization of human neoplasms. Cancer Res. 46, 3183–3191.

Felices, M., Lenvik, T.R., Kodal, B., Lenvik, A.J., Hinderlie, P., Bendzick, L.E., Schirm, D.K., Kaminski, M.F., McElmurry, R.T., Geller, M.A., Eckfeldt, C.E., Vallera, D.A., Miller, J.S., 2020. Potent cytolytic activity and specific IL15 delivery in a second-generation trispecific killer engager. Cancer Immunol. Res. 8, 1139–1149.

Gennari, F., Lopes, L., Verhoeyen, E., Marasco, W., Collins, M.K., 2009. Single-chain antibodies that target lentiviral vectors to MHC class II on antigen-presenting cells. Hum. Gene Ther. 20, 554–562.
Gil, D., Schrum, A.G., 2013. Strategies to stabilize compact folding and minimize aggregation of antibody-based fragments. Adv. Biosci. Biotechnol. 4, 73–84.
Gill, M.R., Falzone, N., Du, Y., Vallis, K.A., 2017. Targeted radionuclide therapy in combined-modality regimens. Lancet Oncol. 18, e414–e423.
Goebeler, M.E., Bargou, R.C., 2020. T cell-engaging therapies—BiTEs and beyond. Nat. Rev. Clin. Oncol. 17, 418–434.
Gong, L., Li, Y., Cui, K., Chen, Y., Hong, H., Li, J., Li, D., Yin, Y., Wu, Z., Huang, Z., 2021. Nanobody-engineered natural killer cell conjugates for solid tumor adoptive immunotherapy. Small 17, e2103463.
Goyvaerts, C., Breckpot, K., 2015. Pros and cons of antigen-presenting cell targeted tumor vaccines. J. Immunol. Res. 2015, 785634.
Goyvaerts, C., De Groeve, K., Dingemans, J., Van Lint, S., Robays, L., Heirman, C., Reiser, J., Zhang, X.Y., Thielemans, K., De Baetselier, P., Raes, G., Breckpot, K., 2012. Development of the nanobody display technology to target lentiviral vectors to antigen-presenting cells. Gene Ther. 19, 1133–1140.
Goyvaerts, C., Dingemans, J., De Groeve, K., Heirman, C., Van Gulck, E., Vanham, G., De Baetselier, P., Thielemans, K., Raes, G., Breckpot, K., 2013. Targeting of human antigen-presenting cell subsets. J. Virol. 87, 11304–11308.
Goyvaerts, C., de Kurt, G., Van Lint, S., Heirman, C., Van Ginderachter, J.A., De Baetselier, P., Raes, G., Thielemans, K., Breckpot, K., 2014. Immunogenicity of targeted lentivectors. Oncotarget 5, 704–715.
Goyvaerts, C., De Vlaeminck, Y., Escors, D., Lienenklaus, S., Keyaerts, M., Raes, G., Breckpot, K., 2017. Antigen-presenting cell-targeted lentiviral vectors do not support the development of productive T-cell effector responses: implications for in vivo targeted vaccine delivery. Gene Ther. 24, 370–375.
Gudkov, S.V., Shilyagina, N.Y., Vodeneev, V.A., Zvyagin, A.V., 2015. Targeted radionuclide therapy of human tumors. Int. J. Mol. Sci. 17, 33.
Guedan, S., Calderon, H., Posey Jr., A.D., Maus, M.V., 2019. Engineering and design of chimeric antigen receptors. Mol. Ther. Methods Clin. Dev. 12, 145–156.
Gul, N., Babes, L., Siegmund, K., Korthouwer, R., Bogels, M., Braster, R., Vidarsson, G., ten Hagen, T.L., Kubes, P., van Egmond, M., 2014. Macrophages eliminate circulating tumor cells after monoclonal antibody therapy. J. Clin. Invest. 124, 812–823.
Gurbatri, C.R., Lia, I., Vincent, R., Coker, C., Castro, S., Treuting, P.M., Hinchliffe, T.E., Arpaia, N., Danino, T., 2020. Engineered probiotics for local tumor delivery of checkpoint blockade nanobodies. Sci. Transl. Med. 12, eaax0876.
Hajari Taheri, F., Hassani, M., Sharifzadeh, Z., Behdani, M., Arashkia, A., Abolhassani, M., 2019. T cell engineered with a novel nanobody-based chimeric antigen receptor against VEGFR2 as a candidate for tumor immunotherapy. IUBMB Life 71, 1259–1267.
Hambach, J., Riecken, K., Cichutek, S., Schutze, K., Albrecht, B., Petry, K., Rockendorf, J.L., Baum, N., Kroger, N., Hansen, T., Schuch, G., Haag, F., Adam, G., Fehse, B., Bannas, P., Koch-Nolte, F., 2020. Targeting CD38-expressing multiple myeloma and Burkitt lymphoma cells in vitro with nanobody-based chimeric antigen receptors (Nb-CARs). Cell 9, 321.
Hamers-Casterman, C., Atarhouch, T., Muyldermans, S., Robinson, G., Hamers, C., Songa, E.B., Bendahman, N., Hamers, R., 1993. Naturally occurring antibodies devoid of light chains. Nature 363, 446–448.
Han, L., Gao, Q., Zhou, K., Zhou, J., Fang, B., Zhang, J., Li, H., Song, Y., 2019. The phase I clinical study of CART targeting BCMA with humanized alpaca-derived single-domain antibody as antigen recognition domain. J. Clin. Oncol. 37, 2535.

Han, L., Zhou, J., Zhou, K., Zhu, X., Zhao, L., Fang, B., Yin, Q., Wei, X., Zhou, H., Li, L., Xu, B., Zhang, J., Song, Y., Gao, Q., 2020. Safety and efficacy of CAR-T cell targeting BCMA in patients with multiple myeloma coinfected with chronic hepatitis B virus. J. Immunother. Cancer 8, e000927.

Hanssens, H., Meeus, F., De Veirman, K., Breckpot, K., Devoogdt, N., 2021. The antigen-binding moiety in the driver's seat of CARs. Med. Res. Rev. 42, 306–342.

Harding, F.A., Stickler, M.M., Razo, J., DuBridge, R.B., 2010. The immunogenicity of humanized and fully human antibodies: residual immunogenicity resides in the CDR regions. MAbs 2, 256–265.

Harwood, S.L., Alvarez-Cienfuegos, A., Nunez-Prado, N., Compte, M., Hernandez-Perez, S., Merino, N., Bonet, J., Navarro, R., Van Bergen En Henegouwen, P.M.P., Lykkemark, S., Mikkelsen, K., Molgaard, K., Jabs, F., Sanz, L., Blanco, F.J., Roda-Navarro, P., Alvarez-Vallina, L., 2017. ATTACK, a novel bispecific T cell-recruiting antibody with trivalent EGFR binding and monovalent CD3 binding for cancer immunotherapy. Oncoimmunology 7, e1377874.

Hassani, M., Hajari Taheri, F., Sharifzadeh, Z., Arashkia, A., Hadjati, J., van Weerden, W.M., Modarressi, M.H., Abolhassani, M., 2019. Construction of a chimeric antigen receptor bearing a nanobody against prostate a specific membrane antigen in prostate cancer. J. Cell. Biochem. 120, 10787–10795.

He, X., Feng, Z., Ma, J., Ling, S., Cao, Y., Gurung, B., Wu, Y., Katona, B.W., O'Dwyer, K.P., Siegel, D.L., June, C.H., Hua, X., 2020. Bispecific and split CAR T cells targeting CD13 and TIM3 eradicate acute myeloid leukemia. Blood 135, 713–723.

Henry, K.A., Hussack, G., Collins, C., Zwaagstra, J.C., Tanha, J., MacKenzie, C.R., 2016. Isolation of TGF-beta-neutralizing single-domain antibodies of predetermined epitope specificity using next-generation DNA sequencing. Protein Eng. Des. Sel. 29, 439–443.

Hernot, S., van Manen, L., Debie, P., Mieog, J.S.D., Vahrmeijer, A.L., 2019. Latest developments in molecular tracers for fluorescence image-guided cancer surgery. Lancet Oncol. 20, e354–e367.

Heukers, R., van Bergen en Henegouwen, P.M., Oliveira, S., 2014. Nanobody-photosensitizer conjugates for targeted photodynamic therapy. Nanomedicine 10, 1441–1451.

Hinni, M.L., Ferlito, A., Brandwein-Gensler, M.S., Takes, R.P., Silver, C.E., Westra, W.H., Seethala, R.R., Rodrigo, J.P., Corry, J., Bradford, C.R., Hunt, J.L., Strojan, P., Devaney, K.O., Gnepp, D.R., Hartl, D.M., Kowalski, L.P., Rinaldo, A., Barnes, L., 2013. Surgical margins in head and neck cancer: a contemporary review. Head Neck 35, 1362–1370.

Hoffmann, R.M., Coumbe, B.G.T., Josephs, D.H., Mele, S., Ilieva, K.M., Cheung, A., Tutt, A.N., Spicer, J.F., Thurston, D.E., Crescioli, S., Karagiannis, S.N., 2018. Antibody structure and engineering considerations for the design and function of Antibody Drug Conjugates (ADCs). Oncoimmunology 7, e1395127.

Homayouni, V., Ganjalikhani-Hakemi, M., Rezaei, A., Khanahmad, H., Behdani, M., Lomedasht, F.K., 2016. Preparation and characterization of a novel nanobody against T-cell immunoglobulin and mucin-3 (TIM-3). Iran. J. Basic Med. Sci. 19, 1201–1208.

Hu, X., Li, B., Li, X., Zhao, X., Wan, L., Lin, G., Yu, M., Wang, J., Jiang, X., Feng, W., Qin, Z., Yin, B., Li, Z., 2014. Transmembrane TNF-alpha promotes suppressive activities of myeloid-derived suppressor cells via TNFR2. J. Immunol. 192, 1320–1331.

Hu, Y., Liu, C., Muyldermans, S., 2017. Nanobody-based delivery systems for diagnosis and targeted tumor therapy. Front. Immunol. 8, 1442.

Huang, R.Y., Francois, A., McGray, A.R., Miliotto, A., Odunsi, K., 2017. Compensatory upregulation of PD-1, LAG-3, and CTLA-4 limits the efficacy of single-agent checkpoint blockade in metastatic ovarian cancer. Oncoimmunology 6, e1249561.

Huet, H.A., Growney, J.D., Johnson, J.A., Li, J., Bilic, S., Ostrom, L., Zafari, M., Kowal, C., Yang, G., Royo, A., Jensen, M., Dombrecht, B., Meerschaert, K.R., Kolkman, J.A., Cromie, K.D., Mosher, R., Gao, H., Schuller, A., Isaacs, R., Sellers, W.R., Ettenberg, S.A., 2014. Multivalent nanobodies targeting death receptor 5 elicit superior tumor cell killing through efficient caspase induction. MAbs 6, 1560–1570.

Huyghe, L., Van Parys, A., Cauwels, A., Van Lint, S., De Munter, S., Bultinck, J., Zabeau, L., Hostens, J., Goethals, A., Vanderroost, N., Verhee, A., Uze, G., Kley, N., Peelman, F., Vandekerckhove, B., Brouckaert, P., Tavernier, J., 2020. Safe eradication of large established tumors using neovasculature-targeted tumor necrosis factor-based therapies. EMBO Mol. Med. 12, e11223.

Inbar, D., Hochman, J., Givol, D., 1972. Localization of antibody-combining sites within the variable portions of heavy and light chains. Proc. Natl. Acad. Sci. U. S. A. 69, 2659–2662.

Ingram, J.R., Blomberg, O.S., Sockolosky, J.T., Ali, L., Schmidt, F.I., Pishesha, N., Espinosa, C., Dougan, S.K., Garcia, K.C., Ploegh, H.L., Dougan, M., 2017a. Localized CD47 blockade enhances immunotherapy for murine melanoma. Proc. Natl. Acad. Sci. U. S. A. 114, 10184–10189.

Ingram, J.R., Dougan, M., Rashidian, M., Knoll, M., Keliher, E.J., Garrett, S., Garforth, S., Blomberg, O.S., Espinosa, C., Bhan, A., Almo, S.C., Weissleder, R., Lodish, H., Dougan, S.K., Ploegh, H.L., 2017b. PD-L1 is an activation-independent marker of brown adipocytes. Nat. Commun. 8, 647.

Ingram, J.R., Blomberg, O.S., Rashidian, M., Ali, L., Garforth, S., Fedorov, E., Fedorov, A.A., Bonanno, J.B., Le Gall, C., Crowley, S., Espinosa, C., Biary, T., Keliher, E.J., Weissleder, R., Almo, S.C., Dougan, S.K., Ploegh, H.L., Dougan, M., 2018. Anti-CTLA-4 therapy requires an Fc domain for efficacy. Proc. Natl. Acad. Sci. U. S. A. 115, 3912–3917.

Iri-Sofla, F.J., Rahbarizadeh, F., Ahmadvand, D., Rasaee, M.J., 2011. Nanobody-based chimeric receptor gene integration in Jurkat cells mediated by phiC31 integrase. Exp. Cell Res. 317, 2630–2641.

Jacoby, E., Shahani, S.A., Shah, N.N., 2019. Updates on CAR T-cell therapy in B-cell malignancies. Immunol. Rev. 290, 39–59.

Jamnani, F.R., Rahbarizadeh, F., Shokrgozar, M.A., Mahboudi, F., Ahmadvand, D., Sharifzadeh, Z., Parhamifar, L., Moghimi, S.M., 2014. T cells expressing VHH-directed oligoclonal chimeric HER2 antigen receptors: towards tumor-directed oligoclonal T cell therapy. Biochim. Biophys. Acta 1840, 378–386.

Ji, X., Peng, Z., Li, X., Yan, Z., Yang, Y., Qiao, Z., Liu, Y., 2017. Neutralization of TNFalpha in tumor with a novel nanobody potentiates paclitaxel-therapy and inhibits metastasis in breast cancer. Cancer Lett. 386, 24–34.

Jou, J., Harrington, K.J., Zocca, M.B., Ehrnrooth, E., Cohen, E.E.W., 2021. The changing landscape of therapeutic cancer vaccines-novel platforms and neoantigen identification. Clin. Cancer Res. 27, 689–703.

Joubert, N., Beck, A., Dumontet, C., Denevault-Sabourin, C., 2020. Antibody-drug conjugates: the last decade. Pharmaceuticals (Basel) 13, 245.

Jovčevska, I., Muyldermans, S., 2020. The therapeutic potential of nanobodies. BioDrugs 34, 11–26.

June, C.H., Sadelain, M., 2018. Chimeric antigen receptor therapy. N. Engl. J. Med. 379, 64–73.

Juweid, M., Neumann, R., Paik, C., Perez-Bacete, M.J., Sato, J., van Osdol, W., Weinstein, J.N., 1992. Micropharmacology of monoclonal antibodies in solid tumors: direct experimental evidence for a binding site barrier. Cancer Res. 52, 5144–5153.

Kang, T.H., Jung, S.T., 2019. Boosting therapeutic potency of antibodies by taming Fc domain functions. Exp. Mol. Med. 51, 1–9.

Kassis, A.I., Adelstein, S.J., 2005. Radiobiologic principles in radionuclide therapy. J. Nucl. Med. 46 (Suppl. 1), 4S–12S.

Khaleghi, S., Rahbarizadeh, F., Ahmadvand, D., Rasaee, M.J., Pognonec, P., 2012. A caspase 8-based suicide switch induces apoptosis in nanobody-directed chimeric receptor expressing T cells. Int. J. Hematol. 95, 434–444.

Khalique, H., Baugh, R., Dyer, A., Scott, E.M., Frost, S., Larkin, S., Lei-Rossmann, J., Seymour, L.W., 2021. Oncolytic herpesvirus expressing PD-L1 BiTE for cancer therapy: exploiting tumor immune suppression as an opportunity for targeted immunotherapy. J. Immunother. Cancer 9, e001292.

Kijanka, M., Warnders, F.J., El Khattabi, M., Lub-de Hooge, M., van Dam, G.M., Ntziachristos, V., de Vries, L., Oliveira, S., van Bergen En Henegouwen, P.M., 2013. Rapid optical imaging of human breast tumour xenografts using anti-HER2 VHHs site-directly conjugated to IRDye 800CW for image-guided surgery. Eur. J. Nucl. Med. Mol. Imaging 40, 1718–1729.

Kijanka, M.M., van Brussel, A.S., van der Wall, E., Mali, W.P., van Diest, P.J., van Bergen En Henegouwen, P.M., Oliveira, S., 2016. Optical imaging of pre-invasive breast cancer with a combination of VHHs targeting CAIX and HER2 increases contrast and facilitates tumour characterization. EJNMMI Res. 6, 14.

Kitamura, Y., Kanaya, N., Moleirinho, S., Du, W., Reinshagen, C., Attia, N., Bronisz, A., Revai Lechtich, E., Sasaki, H., Mora, J.L., Brastianos, P.K., Falcone, J.L., Hofer, A.M., Franco, A., Shah, K., 2021. Anti-EGFR VHH-armed death receptor ligand-engineered allogeneic stem cells have therapeutic efficacy in diverse brain metastatic breast cancers. Sci. Adv. 7, eabe8671.

Krasniqi, A., D'Huyvetter, M., Xavier, C., Van der Jeught, K., Muyldermans, S., Van Der Heyden, J., Lahoutte, T., Tavernier, J., Devoogdt, N., 2017. Theranostic radiolabeled anti-CD20 sdAb for targeted radionuclide therapy of non-Hodgkin lymphoma. Mol. Cancer Ther. 16, 2828–2839.

Lam, J.S., Bergman, J., Breda, A., Schulam, P.G., 2008. Importance of surgical margins in the management of renal cell carcinoma. Nat. Clin. Pract. Urol. 5, 308–317.

Lam, N., Trinklein, N.D., Buelow, B., Patterson, G.H., Ojha, N., Kochenderfer, J.N., 2020. Anti-BCMA chimeric antigen receptors with fully human heavy-chain-only antigen recognition domains. Nat. Commun. 11, 283.

Lauwerends, L.J., van Driel, P.B.A.A., Baatenburg de Jong, R.J., Hardillo, J.A.U., Koljenovic, S., Puppels, G., Mezzanotte, L., Löwik, C.W.G.M., Rosenthal, E.L., Vahrmeijer, A.L., Keereweer, S., 2021. Real-time fluorescence imaging in intraoperative decision making for cancer surgery. Lancet Oncol. 22, e186–e195.

Lecocq, Q., De Vlaeminck, Y., Hanssens, H., D'Huyvetter, M., Raes, G., Goyvaerts, C., Keyaerts, M., Devoogdt, N., Breckpot, K., 2019a. Theranostics in immuno-oncology using nanobody derivatives. Theranostics 9, 7772–7791.

Lecocq, Q., Zeven, K., De Vlaeminck, Y., Martens, S., Massa, S., Goyvaerts, C., Raes, G., Keyaerts, M., Breckpot, K., Devoogdt, N., 2019b. Noninvasive imaging of the immune checkpoint LAG-3 using nanobodies, from development to pre-clinical use. Biomolecules 9.

Lecocq, Q., Keyaerts, M., Devoogdt, N., Breckpot, K., 2020. The next-generation immune checkpoint LAG-3 and its therapeutic potential in oncology: third time's a charm. Int. J. Mol. Sci. 22, 75.

Lecocq, Q., Awad, R.M., De Vlaeminck, Y., de Mey, W., Ertveldt, T., Goyvaerts, C., Raes, G., Thielemans, K., Keyaerts, M., Devoogdt, N., Breckpot, K., 2021a. Single-domain antibody nuclear imaging allows noninvasive quantification of LAG-3 expression by tumor-infiltrating leukocytes and predicts response of immune checkpoint blockade. J. Nucl. Med. 62, 1638–1644.

Lecocq, Q., Debie, P., Puttemans, J., Awad, R.M., De Beck, L., Ertveldt, T., De Vlaeminck, Y., Goyvaerts, C., Raes, G., Keyaerts, M., Breckpot, K., Devoogdt, N., 2021b. Evaluation of single domain antibodies as nuclear tracers for imaging of the immune checkpoint receptor human lymphocyte activation gene-3 in cancer. EJNMMI Res. 11, 115.

Lemaire, M., D'Huyvetter, M., Lahoutte, T., Van Valckenborgh, E., Menu, E., De Bruyne, E., Kronenberger, P., Wernery, U., Muyldermans, S., Devoogdt, N., Vanderkerken, K., 2014. Imaging and radioimmunotherapy of multiple myeloma with anti-idiotypic nanobodies. Leukemia 28, 444–447.

Li, L., He, P., Zhou, C., Jing, L., Dong, B., Chen, S., Zhang, N., Liu, Y., Miao, J., Wang, Z., Li, Q., 2015. A novel bispecific antibody, S-Fab, induces potent cancer cell killing. J. Immunother. 38, 350–356.

Li, A., Xing, J., Li, L., Zhou, C., Dong, B., He, P., Li, Q., Wang, Z., 2016a. A single-domain antibody-linked Fab bispecific antibody Her2-S-Fab has potent cytotoxicity against Her2-expressing tumor cells. AMB Express 6, 32.

Li, J., Zhou, C., Dong, B., Zhong, H., Chen, S., Li, Q., Wang, Z., 2016b. Single domain antibody-based bispecific antibody induces potent specific anti-tumor activity. Cancer Biol. Ther. 17, 1231–1239.

Li, N., Fu, H., Hewitt, S.M., Dimitrov, D.S., Ho, M., 2017. Therapeutically targeting glypican-2 via single-domain antibody-based chimeric antigen receptors and immunotoxins in neuroblastoma. Proc. Natl. Acad. Sci. U. S. A. 114, E6623–E6631.

Li, Y., Zhou, C., Li, J., Liu, J., Lin, L., Li, L., Cao, D., Li, Q., Wang, Z., 2018. Single domain based bispecific antibody, Muc1-Bi-1, and its humanized form, Muc1-Bi-2, induce potent cancer cell killing in muc1 positive tumor cells. PLoS One 13, e0191024.

Li, J., Deng, Y., Zhang, W., Zhou, A.P., Guo, W., Yang, J., Yuan, Y., Zhu, L., Qin, S., Xiang, S., Lu, H., Gong, J., Xu, T., Liu, D., Shen, L., 2021. Subcutaneous envafolimab monotherapy in patients with advanced defective mismatch repair/microsatellite instability high solid tumors. J. Hematol. Oncol. 14, 95.

Liu, Y., Wang, Y., Xing, J., Li, Y., Liu, J., Wang, Z., 2018. A novel multifunctional anti-CEA-IL15 molecule displays potent antitumor activities. Drug Des. Devel. Ther. 12, 2645–2654.

Liu, Y., Ji, X., Kang, N., Zhou, J., Liang, X., Li, J., Han, T., Zhao, C., Yang, T., 2020. Tumor necrosis factor alpha inhibition overcomes immunosuppressive M2b macrophage-induced bevacizumab resistance in triple-negative breast cancer. Cell Death Dis. 11, 993.

Long, A.H., Haso, W.M., Shern, J.F., Wanhainen, K.M., Murgai, M., Ingaramo, M., Smith, J.P., Walker, A.J., Kohler, M.E., Venkateshwara, V.R., Kaplan, R.N., Patterson, G.H., Fry, T.J., Orentas, R.J., Mackall, C.L., 2015. 4-1BB costimulation ameliorates T cell exhaustion induced by tonic signaling of chimeric antigen receptors. Nat. Med. 21, 581–590.

Lucas, A.T., Price, L.S.L., Schorzman, A.N., Storrie, M., Piscitelli, J.A., Razo, J., Zamboni, W.C., 2018. Factors affecting the pharmacology of antibody-drug conjugates. Antibodies (Basel) 7, 10.

Lwin, T.M., Hernot, S., Hollandsworth, H., Amirfakhri, S., Filemoni, F., Debie, P., Hoffman, R.M., Bouvet, M., 2020. Tumor-specific near-infrared nanobody probe rapidly labels tumors in an orthotopic mouse model of pancreatic cancer. Surgery 168, 85–91.

Lwin, T.M., Turner, M.A., Amirfakhri, S., Nishino, H., Debie, P., Cosman, B.C., Hoffman, R.M., Hernot, S., Bouvet, M., 2021. Rapid tumor-labeling kinetics with a site-specific near-infrared anti-CEA nanobody in a patient-derived orthotopic xenograft mouse model of colon cancer. J. Surg. Oncol. 124, 1121–1127.

Ma, L., Gai, J., Qiao, P., Li, Y., Li, X., Zhu, M., Li, G., Wan, Y., 2020a. A novel bispecific nanobody with PD-L1/TIGIT dual immune checkpoint blockade. Biochem. Biophys. Res. Commun. 531, 144–151.

Ma, L., Zhu, M., Gai, J., Li, G., Chang, Q., Qiao, P., Cao, L., Chen, W., Zhang, S., Wan, Y., 2020b. Preclinical development of a novel CD47 nanobody with less toxicity and enhanced anti-cancer therapeutic potential. J. Nanobiotechnology 18, 12.

Macri, C., Dumont, C., Johnston, A.P., Mintern, J.D., 2016. Targeting dendritic cells: a promising strategy to improve vaccine effectiveness. Clin. Transl. Immunol. 5, e66.

Madduri, D., Berdeja, J.G., Usmani, S.Z., Jakubowiak, A., Agha, M., Cohen, A.D., Stewart, A.K., Hari, P., Htut, M., O'Donnell, E., Munshi, N.C., Avigan, D.E., Deol, A., Lesokhin, A.M., Singh, I., Zudaire, E., Yeh, T.-M., Allred, A.J., Olyslager, Y., Banerjee, A., Goldberg, J.D., Schecter, J.M., Jackson, C.C., Deraedt, W., Zhuang, S.H., Infante, J.R., Geng, D., Wu, X., Carrasco, M.J., Akram, M., Hossain, F., Rizvi, S., Fan, F., Jagannath, S., Lin, Y., Martin, III, T., 2020. CARTITUDE-1: phase 1b/2 study of ciltacabtagene autoleucel, a B-cell maturation antigen-directed chimeric antigen receptor T cell therapy, in relapsed/refractory multiple myeloma. Blood 136, 22–25.

Marcucci, F., Bellone, M., Rumio, C., Corti, A., 2013. Approaches to improve tumor accumulation and interactions between monoclonal antibodies and immune cells. MAbs 5, 34–46.

Massa, S., Xavier, C., De Vos, J., Caveliers, V., Lahoutte, T., Muyldermans, S., Devoogdt, N., 2014. Site-specific labeling of cysteine-tagged camelid single-domain antibody-fragments for use in molecular imaging. Bioconjug. Chem. 25, 979–988.

Massa, S., Vikani, N., Betti, C., Ballet, S., Vanderhaegen, S., Steyaert, J., Descamps, B., Vanhove, C., Bunschoten, A., van Leeuwen, F.W., Hernot, S., Caveliers, V., Lahoutte, T., Muyldermans, S., Xavier, C., Devoogdt, N., 2016. Sortase A-mediated site-specific labeling of camelid single-domain antibody-fragments: a versatile strategy for multiple molecular imaging modalities. Contrast Media Mol. Imaging 11, 328–339.

Mikkelsen, K., Harwood, S.L., Compte, M., Merino, N., Molgaard, K., Lykkemark, S., Alvarez-Mendez, A., Blanco, F.J., Alvarez-Vallina, L., 2019. Carcinoembryonic antigen (CEA)-specific 4-1BB-costimulation induced by CEA-targeted 4-1BB-agonistic trimerbodies. Front. Immunol. 10, 1791.

Minetto, P., Guolo, F., Pesce, S., Greppi, M., Obino, V., Ferretti, E., Sivori, S., Genova, C., Lemoli, R.M., Marcenaro, E., 2019. Harnessing NK cells for cancer treatment. Front. Immunol. 10, 2836.

Miyashita, M., Shimizu, T., Ashihara, E., Ukimura, O., 2021. Strategies to improve the anti-tumor effect of γδ T cell immunotherapy for clinical application. Int. J. Mol. Sci. 22, 8910.

Moeller, E., 1965. Contact-induced cytotoxicity by lymphoid cells containing foreign isoantigens. Science (New York, N.Y.) 147, 873–879.

Molgaard, K., Harwood, S.L., Compte, M., Merino, N., Bonet, J., Alvarez-Cienfuegos, A., Mikkelsen, K., Nunez-Prado, N., Alvarez-Mendez, A., Sanz, L., Blanco, F.J., Alvarez-Vallina, L., 2018. Bispecific light T-cell engagers for gene-based immunotherapy of epidermal growth factor receptor (EGFR)-positive malignancies. Cancer Immunol. Immunother. 67, 1251–1260.

Montfort, A., Colacios, C., Levade, T., Andrieu-Abadie, N., Meyer, N., Segui, B., 2019. The TNF paradox in cancer progression and immunotherapy. Front. Immunol. 10, 1818.

Moradian, C., Rahbarizadeh, F., 2021. PE38-based gene therapy of HER2-positive breast cancer stem cells via VHH-redirected polyamidoamine dendrimers. Sci. Rep. 11, 15517.

Movahedi, K., Schoonooghe, S., Laoui, D., Houbracken, I., Waelput, W., Breckpot, K., Bouwens, L., Lahoutte, T., De Baetselier, P., Raes, G., Devoogdt, N., Van Ginderachter, J.A., 2012. Nanobody-based targeting of the macrophage mannose receptor for effective in vivo imaging of tumor-associated macrophages. Cancer Res. 72, 4165–4177.

Muller, B.G., Leuenberger, H., Kissel, T., 1996. Albumin nanospheres as carriers for passive drug targeting: an optimized manufacturing technique. Pharm. Res. 13, 32–37.

Muyldermans, S., 2013. Nanobodies: natural single-domain antibodies. Annu. Rev. Biochem. 82, 775–797.

Nagaya, T., Nakamura, Y.A., Choyke, P.L., Kobayashi, H., 2017. Fluorescence-guided surgery. Front. Oncol. 7, 314.

Naidoo, J., Page, D.B., Li, B.T., Connell, L.C., Schindler, K., Lacouture, M.E., Postow, M.A., Wolchok, J.D., 2015. Toxicities of the anti-PD-1 and anti-PD-L1 immune checkpoint antibodies. Ann. Oncol. 26, 2375–2391.

O'Kelly Priddy, C.M., Forte, V.A., Lang, J.E., 2016. The importance of surgical margins in breast cancer. J. Surg. Oncol. 113, 256–263.

Oliveira, S., Schiffelers, R.M., van der Veeken, J., van der Meel, R., Vongpromek, R., van Bergen En Henegouwen, P.M., Storm, G., Roovers, R.C., 2010. Downregulation of EGFR by a novel multivalent nanobody-liposome platform. J. Control. Release 145, 165–175.

Oliveira, S., van Dongen, G.A., Stigter-van Walsum, M., Roovers, R.C., Stam, J.C., Mali, W., van Diest, P.J., van Bergen en Henegouwen, P.M., 2012. Rapid visualization of human tumor xenografts through optical imaging with a near-infrared fluorescent anti-epidermal growth factor receptor nanobody. Mol. Imaging 11, 33–46.

Orosco, R.K., Tapia, V.J., Califano, J.A., Clary, B., Cohen, E.E.W., Kane, C., Lippman, S.M., Messer, K., Molinolo, A., Murphy, J.D., Pang, J., Sacco, A., Tringale, K.R., Wallace, A., Nguyen, Q.T., 2018. Positive surgical margins in the 10 most common solid cancers. Sci. Rep. 8, 5686.

Pan, H., Liu, J., Deng, W., Xing, J., Li, Q., Wang, Z., 2018. Site-specific PEGylation of an anti-CEA/CD3 bispecific antibody improves its antitumor efficacy. Int. J. Nanomedicine 13, 3189–3201.

Papadopoulos, K.P., Harb, W., Peer, C.J., Hua, Q., Xu, S., Lu, H., Lu, N., He, Y., Xu, T., Dong, R., Gong, J., Liu, D., 2021. First-in-human phase I study of envafolimab, a novel subcutaneous single-domain anti-PD-L1 antibody, in patients with advanced solid tumors. Oncologist 26, e1514–e1525.

Papaioannou, N.E., Beniata, O.V., Vitsos, P., Tsitsilonis, O., Samara, P., 2016. Harnessing the immune system to improve cancer therapy. Ann. Transl. Med. 4, 261.

Papp, K.A., Weinberg, M.A., Morris, A., Reich, K., 2021. IL17A/F nanobody sonelokimab in patients with plaque psoriasis: a multicentre, randomised, placebo-controlled, phase 2b study. Lancet 397, 1564–1575.

Paston, S.J., Brentville, V.A., Symonds, P., Durrant, L.G., 2021. Cancer vaccines, adjuvants, and delivery systems. Front. Immunol. 12, 627932.

Patra, J.K., Das, G., Fraceto, L.F., Campos, E.V.R., Rodriguez-Torres, M.D.P., Acosta-Torres, L.S., Diaz-Torres, L.A., Grillo, R., Swamy, M.K., Sharma, S., Habtemariam, S., Shin, H.S., 2018. Nano based drug delivery systems: recent developments and future prospects. J. Nanobiotechnol. 16, 71.

Pauling, L., 1940. A theory of the structure and process of formation of antibodies. J. Am. Chem. Soc. 62, 2643–2657.

Pishesha, N., Harmand, T., Smeding, L.Y., Ma, W., Ludwig, L.S., Janssen, R., Islam, A., Xie, Y.J., Fang, T., McCaul, N., Pinney, W., 3rd, Sugito, H.R., Rossotti, M.A., Gonzalez-Sapienza, G., Ploegh, H.L., 2021. Induction of antigen-specific tolerance by nanobody-antigen adducts that target class-II major histocompatibility complexes. Nat. Biomed. Eng. 5, 1389–1401.

Pleiner, T., Bates, M., Trakhanov, S., Lee, C.T., Schliep, J.E., Chug, H., Böhning, M., Stark, H., Urlaub, H., Görlich, D., 2015. Nanobodies: site-specific labeling for super-resolution imaging, rapid epitope-mapping and native protein complex isolation. Elife 4, e11349.

Porter, R.R., 1959. The hydrolysis of rabbit y-globulin and antibodies with crystalline papain. Biochem. J. 73, 119–126.

Pouget, J.P., Navarro-Teulon, I., Bardies, M., Chouin, N., Cartron, G., Pelegrin, A., Azria, D., 2011. Clinical radioimmunotherapy—the role of radiobiology. Nat. Rev. Clin. Oncol. 8, 720–734.

Pruszynski, M., D'Huyvetter, M., Bruchertseifer, F., Morgenstern, A., Lahoutte, T., 2018. Evaluation of an anti-HER2 nanobody labeled with (225)Ac for targeted alpha-particle therapy of cancer. Mol. Pharm. 15, 1457–1466.

Puttemans, J., Dekempeneer, Y., Eersels, J.L., Hanssens, H., Debie, P., Keyaerts, M., Windhorst, A.D., van der Aa, F., Lecocq, Q., Breckpot, K., Morgenstern, A., Bruchertseifer, F., Lahoutte, T., Devoogdt, N., D'Huyvetter, M., 2020. Preclinical targeted alpha- and beta(-)-radionuclide therapy in HER2-positive brain metastasis using camelid single-domain antibodies. Cancers (Basel) 12.

Puttemans, J., Stijlemans, B., Keyaerts, M., Vander Meeren, S., Renmans, W., Fostier, K., Debie, P., Hanssens, H., Rodak, M., Pruszynski, M., De Veirman, K., Vanderkerken, K., Lahoutte, T., Morgenstern, A., Bruchertseifer, F., Devoogdt, N., D'Huyvetter, M., 2021. The road to personalized myeloma medicine: patient-specific single-domain antibodies for anti-idiotypic radionuclide therapy. Mol. Cancer Ther. 21, 159–169.

Rafiq, S., Hackett, C.S., Brentjens, R.J., 2020. Engineering strategies to overcome the current roadblocks in CAR T cell therapy. Nat. Rev. Clin. Oncol. 17, 147–167.

Rahbarizadeh, F., Ahmadvand, D., Moghimi, S.M., 2019. CAR T-cell bioengineering: single variable domain of heavy chain antibody targeted CARs. Adv. Drug Deliv. Rev. 141, 41–46.

Renard, E., Collado Camps, E., Canovas, C., Kip, A., Gotthardt, M., Rijpkema, M., Denat, F., Goncalves, V., van Lith, S.A.M., 2021. Site-specific dual-labeling of a VHH with a chelator and a photosensitizer for nuclear imaging and targeted photodynamic therapy of EGFR positive tumors. Cancers (Basel) 13.

Rouard, H., Tamasdan, S., Moncuit, J., Moutel, S., Michon, J., Fridman, W.H., Teillaud, J.L., 1997. Fc receptors as targets for immunotherapy. Int. Rev. Immunol. 16, 147–185.

Rozan, C., Cornillon, A., Petiard, C., Chartier, M., Behar, G., Boix, C., Kerfelec, B., Robert, B., Pelegrin, A., Chames, P., Teillaud, J.L., Baty, D., 2013. Single-domain antibody-based and linker-free bispecific antibodies targeting FcgammaRIII induce potent antitumor activity without recruiting regulatory T cells. Mol. Cancer Ther. 12, 1481–1491.

Sadeghnezhad, G., Romao, E., Bernedo-Navarro, R., Massa, S., Khajeh, K., Muyldermans, S., Hassania, S., 2019. Identification of new DR5 agonistic nanobodies and generation of multivalent nanobody constructs for cancer treatment. Int. J. Mol. Sci. 20, 4818.

Sadelain, M., 2017. Chimeric antigen receptors: a paradigm shift in immunotherapy. Annu. Rev. Cancer Biol. 1, 447–466.

Sadelain, M., Brentjens, R., Riviere, I., 2013. The basic principles of chimeric antigen receptor design. Cancer Discov. 3, 388–398.

Scott, A.M., Wolchok, J.D., Old, L.J., 2012. Antibody therapy of cancer. Nat. Rev. Cancer 12, 278–287.

Sharifzadeh, Z., Rahbarizadeh, F., Shokrgozar, M.A., Ahmadvand, D., Mahboudi, F., Jamnani, F.R., Moghimi, S.M., 2013. Genetically engineered T cells bearing chimeric nanoconstructed receptors harboring TAG-72-specific camelid single domain antibodies as targeting agents. Cancer Lett. 334, 237–244.

Sharma, P.K., Dmitriev, I.P., Kashentseva, E.A., Raes, G., Li, L., Kim, S.W., Lu, Z.H., Arbeit, J.M., Fleming, T.P., Kaliberov, S.A., Goedegebuure, S.P., Curiel, D.T., Gillanders, W.E., 2018. Development of an adenovirus vector vaccine platform for targeting dendritic cells. Cancer Gene Ther. 25, 27–38.

Sievers, N.M., Dorrie, J., Schaft, N., 2020. CARs: beyond T cells and T cell-derived signaling domains. Int. J. Mol. Sci. 21, 3525.

Silva-Pilipich, N., Martisova, E., Ballesteros-Briones, M.C., Hervas-Stubbs, S., Casares, N., Gonzalez-Sapienza, G., Smerdou, C., Vanrell, L., 2020. Long-term systemic expression of a novel PD-1 blocking nanobody from an AAV vector provides antitumor activity without toxicity. Biomedicines 8, 562.

Sockolosky, J.T., Dougan, M., Ingram, J.R., Ho, C.C., Kauke, M.J., Almo, S.C., Ploegh, H.L., Garcia, K.C., 2016. Durable antitumor responses to CD47 blockade require adaptive immune stimulation. Proc. Natl. Acad. Sci. U. S. A. 113, E2646–E2654.

Syed, V., 2016. TGF-beta signaling in cancer. J. Cell. Biochem. 117, 1279–1287.

Talelli, M., Rijcken, C.J., Oliveira, S., van der Meel, R., van Bergen En Henegouwen, P.M., Lammers, T., van Nostrum, C.F., Storm, G., Hennink, W.E., 2011. Nanobody-shell functionalized thermosensitive core-crosslinked polymeric micelles for active drug targeting. J. Control. Release 151, 183–192.

Talelli, M., Oliveira, S., Rijcken, C.J., Pieters, E.H., Etrych, T., Ulbrich, K., van Nostrum, R.C., Storm, G., Hennink, W.E., Lammers, T., 2013. Intrinsically active nanobody-modified polymeric micelles for tumor-targeted combination therapy. Biomaterials 34, 1255–1260.

Tang, Z., Mo, F., Liu, A., Duan, S., Yang, X., Liang, L., Hou, X., Yin, S., Jiang, X., Vasylieva, N., Dong, J., Barnych, B., Hammock, B.D., Lu, X., 2019. A nanobody against cytotoxic T-lymphocyte associated antigen-4 increases the anti-tumor effects of specific CD8(+) T cells. J. Biomed. Nanotechnol. 15, 2229–2239.

Teng, M.W., Andrews, D.M., McLaughlin, N., von Scheidt, B., Ngiow, S.F., Moller, A., Hill, G.R., Iwakura, Y., Oft, M., Smyth, M.J., 2010. IL-23 suppresses innate immune response independently of IL-17A during carcinogenesis and metastasis. Proc. Natl. Acad. Sci. U. S. A. 107, 8328–8333.

Teng, M.W., von Scheidt, B., Duret, H., Towne, J.E., Smyth, M.J., 2011. Anti-IL-23 monoclonal antibody synergizes in combination with targeted therapies or IL-2 to suppress tumor growth and metastases. Cancer Res. 71, 2077–2086.

Toffoli, E.C., Sheikhi, A., Lameris, R., King, L.A., van Vliet, A., Walcheck, B., Verheul, H.M.W., Spanholtz, J., Tuynman, J., de Gruijl, T.D., van der Vliet, H.J., 2021. Enhancement of NK cell antitumor effector functions using a bispecific single domain antibody targeting CD16 and the epidermal growth factor receptor. Cancers (Basel) 13, 5446.

Turini, M., Chames, P., Bruhns, P., Baty, D., Kerfelec, B., 2014. A FcgammaRIII-engaging bispecific antibody expands the range of HER2-expressing breast tumors eligible to antibody therapy. Oncotarget 5, 5304–5319.

Vallera, D.A., Ferrone, S., Kodal, B., Hinderlie, P., Bendzick, L., Ettestad, B., Hallstrom, C., Zorko, N.A., Rao, A., Fujioka, N., Ryan, C.J., Geller, M.A., Miller, J.S., Felices, M., 2020. NK-cell-mediated targeting of various solid tumors using a B7-H3 tri-specific killer engager in vitro and in vivo. Cancers (Basel) 12, 2659.

Vallera, D.A., Oh, F., Kodal, B., Hinderlie, P., Geller, M.A., Miller, J.S., Felices, M., 2021. A HER2 tri-specific NK cell engager mediates efficient targeting of human ovarian cancer. Cancers (Basel) 13, 3994.

van Brussel, A.S., Adams, A., Oliveira, S., Dorresteijn, B., El Khattabi, M., Vermeulen, J.F., van der Wall, E., Mali, W.P., Derksen, P.W., van Diest, P.J., van Bergen En Henegouwen, P.M., 2016. Hypoxia-targeting fluorescent nanobodies for optical molecular imaging of pre-invasive breast cancer. Mol. Imaging Biol. 18, 535–544.

Van Damme, H., Dombrecht, B., Kiss, M., Roose, H., Allen, E., Van Overmeire, E., Kancheva, D., Martens, L., Murgaski, A., Bardet, P.M.R., Blancke, G., Jans, M., Bolli, E., Martins, M.S., Elkrim, Y., Dooley, J., Boon, L., Schwarze, J.K., Tacke, F., Movahedi, K., Vandamme, N., Neyns, B., Ocak, S., Scheyltjens, I., Vereecke, L., Nana, F.A., Merchiers, P., Laoui, D., Van Ginderachter, J.A., 2021. Therapeutic depletion of CCR8(+) tumor-infiltrating regulatory T cells elicits antitumor immunity and synergizes with anti-PD-1 therapy. J. Immunother. Cancer 9, e001749.

Van Den Eeckhout, B., Van Hoecke, L., Burg, E., Van Lint, S., Peelman, F., Kley, N., Uze, G., Saelens, X., Tavernier, J., Gerlo, S., 2020. Specific targeting of IL-1beta activity to CD8(+) T cells allows for safe use as a vaccine adjuvant. NPJ Vaccines 5, 64.

Van Den Eeckhout, B., Huyghe, L., Van Lint, S., Burg, E., Plaisance, S., Peelman, F., Cauwels, A., Uze, G., Kley, N., Gerlo, S., Tavernier, J., 2021. Selective IL-1 activity on CD8(+) T cells empowers antitumor immunity and synergizes with neovasculature-targeted TNF for full tumor eradication. J. Immunother. Cancer 9, e003293.

van der Bruggen, P., Traversari, C., Chomez, P., Lurquin, C., De Plaen, E., Van den Eynde, B., Knuth, A., Boon, T., 1991. A gene encoding an antigen recognized by cytolytic T lymphocytes on a human melanoma. Science (New York, N.Y.) 254, 1643–1647.

van der Meel, R., Oliveira, S., Altintas, I., Haselberg, R., van der Veeken, J., Roovers, R.C., van Bergen en Henegouwen, P.M., Storm, G., Hennink, W.E., Schiffelers, R.M., Kok, R.J., 2012. Tumor-targeted nanobullets: anti-EGFR nanobody-liposomes loaded with anti-IGF-1R kinase inhibitor for cancer treatment. J. Control. Release 159, 281–289.

van der Meel, R., Oliveira, S., Altintas, I., Heukers, R., Pieters, E.H., van Bergen en Henegouwen, P.M., Storm, G., Hennink, W.E., Kok, R.J., Schiffelers, R.M., 2013. Inhibition of tumor growth by targeted anti-EGFR/IGF-1R nanobullets depends on efficient blocking of cell survival pathways. Mol. Pharm. 10, 3717–3727.

van de Water, J.A., Bagci-Onder, T., Agarwal, A.S., Wakimoto, H., Roovers, R.C., Zhu, Y., Kasmieh, R., Bhere, D., Van Bergen en Henegouwen, P.M., Shah, K., 2012. Therapeutic stem cells expressing variants of EGFR-specific nanobodies have antitumor effects. Proc. Natl. Acad. Sci. U. S. A. 109, 16642–16647.

van Driel, P.B., van der Vorst, J.R., Verbeek, F.P., Oliveira, S., Snoeks, T.J., Keereweer, S., Chan, B., Boonstra, M.C., Frangioni, J.V., van Bergen en Henegouwen, P.M., Vahrmeijer, A.L., Lowik, C.W., 2014. Intraoperative fluorescence delineation of head and neck cancer with a fluorescent anti-epidermal growth factor receptor nanobody. Int. J. Cancer 134, 2663–2673.

van Driel, P., Boonstra, M.C., Slooter, M.D., Heukers, R., Stammes, M.A., Snoeks, T.J.A., de Bruijn, H.S., van Diest, P.J., Vahrmeijer, A.L., van Bergen En Henegouwen, P.M.P., van de Velde, C.J.H., Lowik, C., Robinson, D.J., Oliveira, S., 2016. EGFR targeted nanobody-photosensitizer conjugates for photodynamic therapy in a pre-clinical model of head and neck cancer. J. Control. Release 229, 93–105.

van Faassen, H., Jo, D.H., Ryan, S., Lowden, M.J., Raphael, S., MacKenzie, C.R., Lee, S.H., Hussack, G., Henry, K.A., 2021. Incorporation of a novel CD16-specific single-domain antibody into multispecific natural killer cell engagers with potent ADCC. Mol. Pharm. 18, 2375–2384.

van Leeuwen, F.W., Valdés-Olmos, R., Buckle, T., Vidal-Sicart, S., 2016. Hybrid surgical guidance based on the integration of radionuclear and optical technologies. Br. J. Radiol. 89, 20150797.

Vaneycken, I., Devoogdt, N., Van Gassen, N., Vincke, C., Xavier, C., Wernery, U., Muyldermans, S., Lahoutte, T., Caveliers, V., 2011. Preclinical screening of anti-HER2 nanobodies for molecular imaging of breast cancer. FASEB J. 25, 2433–2446.

Verhaar, E.R., Woodham, A.W., Ploegh, H.L., 2021. Nanobodies in cancer. Semin. Immunol. 52, 101425.
Vermaelen, K., 2019. Vaccine strategies to improve anti-cancer cellular immune responses. Front. Immunol. 10, 8.
Vincke, C., Loris, R., Saerens, D., Martinez-Rodriguez, S., Muyldermans, S., Conrath, K., 2009. General strategy to humanize a camelid single-domain antibody and identification of a universal humanized nanobody scaffold. J. Biol. Chem. 284, 3273–3284.
Wan, R., Liu, A., Hou, X., Lai, Z., Li, J., Yang, N., Tan, J., Mo, F., Hu, Z., Yang, X., Zhao, Y., Lu, X., 2018. Screening and antitumor effect of an antiCTLA4 nanobody. Oncol. Rep. 39, 511–518.
Wang, Y., Liu, J., Pan, H., Xing, J., Wu, X., Li, Q., Wang, Z., 2018. A GPC3-targeting bispecific antibody, GPC3-S-Fab, with potent cytotoxicity. J. Vis. Exp. (137), 57588.
Wang, H., Wang, L., Li, Y., Li, G., Zhang, X., Jiang, D., Zhang, Y., Liu, L., Chu, Y., Xu, G., 2021. Nanobody-armed T cells endow CAR-T cells with cytotoxicity against lymphoma cells. Cancer Cell Int. 21, 450.
Wen, B., Zhao, L., Wang, Y., Qiu, C., Xu, Z., Huang, K., Zhu, H., Li, Z., Li, H., 2020. Nanobodies targeting the interaction interface of programmed death receptor 1 (PD-1)/PD-1 ligand 1 (PD-1/PD-L1). Prep. Biochem. Biotechnol. 50, 252–259.
Wirth, T.C., Kuhnel, F., 2017. Neoantigen targeting-dawn of a new era in cancer immunotherapy? Front. Immunol. 8, 1848.
Witzig, T.E., White, C.A., Gordon, L.I., Wiseman, G.A., Emmanouilides, C., Murray, J.L., Lister, J., Multani, P.S., 2003. Safety of yttrium-90 ibritumomab tiuxetan radioimmunotherapy for relapsed low-grade, follicular, or transformed non-Hodgkin's lymphoma. J. Clin. Oncol. 21, 1263–1270.
Woodham, A.W., Cheloha, R.W., Ling, J., Rashidian, M., Kolifrath, S.C., Mesyngier, M., Duarte, J.N., Bader, J.M., Skeate, J.G., Da Silva, D.M., Kast, W.M., Ploegh, H.L., 2018. Nanobody-antigen conjugates elicit HPV-specific antitumor immune responses. Cancer Immunol. Res. 6, 870–880.
Wu, T., Liu, M., Huang, H., Sheng, Y., Xiao, H., Liu, Y., 2020. Clustered nanobody-drug conjugates for targeted cancer therapy. Chem. Commun. (Camb.) 56, 9344–9347.
Xian, Z., Ma, L., Zhu, M., Li, G., Gai, J., Chang, Q., Huang, Y., Ju, D., Wan, Y., 2019. Blocking the PD-1-PD-L1 axis by a novel PD-1 specific nanobody expressed in yeast as a potential therapeutic for immunotherapy. Biochem. Biophys. Res. Commun. 519, 267–273.
Xie, Y.J., Dougan, M., Jailkhani, N., Ingram, J., Fang, T., Kummer, L., Momin, N., Pishesha, N., Rickelt, S., Hynes, R.O., Ploegh, H., 2019. Nanobody-based CAR T cells that target the tumor microenvironment inhibit the growth of solid tumors in immunocompetent mice. Proc. Natl. Acad. Sci. U. S. A. 116, 7624–7631.
Xie, Y.J., Dougan, M., Ingram, J.R., Pishesha, N., Fang, T., Momin, N., Ploegh, H.L., 2020. Improved antitumor efficacy of chimeric antigen receptor T cells that secrete single-domain antibody fragments. Cancer Immunol. Res. 8, 518–529.
Xu, J., Chen, L.J., Yang, S.S., Sun, Y., Wu, W., Liu, Y.F., Xu, J., Zhuang, Y., Zhang, W., Weng, X.Q., Wu, J., Wang, Y., Wang, J., Yan, H., Xu, W.B., Jiang, H., Du, J., Ding, X.Y., Li, B., Li, J.M., Fu, W.J., Zhu, J., Zhu, L., Chen, Z., Fan, X.F., Hou, J., Li, J.Y., Mi, J.Q., Chen, S.J., 2019. Exploratory trial of a biepitopic CAR T-targeting B cell maturation antigen in relapsed/refractory multiple myeloma. Proc. Natl. Acad. Sci. U. S. A. 116, 9543–9551.
Yan, J., Smyth, M.J., Teng, M.W.L., 2018. Interleukin (IL)-12 and IL-23 and their conflicting roles in cancer. Cold Spring Harb. Perspect. Biol. 10, a028530.
Yang, E.Y., Shah, K., 2020. Nanobodies: next generation of cancer diagnostics and therapeutics. Front. Oncol. 10, 1182.

Yang, C., Gao, X., Gong, R., 2017. Engineering of Fc fragments with optimized physicochemical properties implying improvement of clinical potentials for Fc-based therapeutics. Front. Immunol. 8, 1860.
Yossepowitch, O., Briganti, A., Eastham, J.A., Epstein, J., Graefen, M., Montironi, R., Touijer, K., 2014. Positive surgical margins after radical prostatectomy: a systematic review and contemporary update. Eur. Urol. 65, 303–313.
You, F., Wang, Y., Jiang, L., Zhu, X., Chen, D., Yuan, L., An, G., Meng, H., Yang, L., 2019. A novel CD7 chimeric antigen receptor-modified NK-92MI cell line targeting T-cell acute lymphoblastic leukemia. Am. J. Cancer Res. 9, 64–78.
Yu, Y., Li, J., Zhu, X., Tang, X., Bao, Y., Sun, X., Huang, Y., Tian, F., Liu, X., Yang, L., 2017. Humanized CD7 nanobody-based immunotoxins exhibit promising anti-T-cell acute lymphoblastic leukemia potential. Int. J. Nanomedicine 12, 1969–1983.
Zahavi, D., Weiner, L., 2020. Monoclonal antibodies in cancer therapy. Antibodies (Basel) 9, 34.
Zeng, B., Middelberg, A.P., Gemiarto, A., MacDonald, K., Baxter, A.G., Talekar, M., Moi, D., Tullett, K.M., Caminschi, I., Lahoud, M.H., Mazzieri, R., Dolcetti, R., Thomas, R., 2018. Self-adjuvanting nanoemulsion targeting dendritic cell receptor Clec9A enables antigen-specific immunotherapy. J. Clin. Invest. 128, 1971–1984.
Zhai, T., Wang, C., Xu, Y., Huang, W., Yuan, Z., Wang, T., Dai, S., Peng, S., Pang, T., Jiang, W., Huang, Y., Zou, Y., Xu, Y., Sun, J., Gong, X., Zhang, J., Tsun, A., Li, B., Miao, X., 2021. Generation of a safe and efficacious llama single-domain antibody fragment (vHH) targeting the membrane-proximal region of 4-1BB for engineering therapeutic bispecific antibodies for cancer. J. Immunother. Cancer 9.
Zhang, G., Wang, L., Cui, H., Wang, X., Zhang, G., Ma, J., Han, H., He, W., Wang, W., Zhao, Y., Liu, C., Sun, M., Gao, B., 2014. Anti-melanoma activity of T cells redirected with a TCR-like chimeric antigen receptor. Sci. Rep. 4, 3571.
Zhang, F., Wei, H., Wang, X., Bai, Y., Wang, P., Wu, J., Jiang, X., Wang, Y., Cai, H., Xu, T., Zhou, A., 2017. Structural basis of a novel PD-L1 nanobody for immune checkpoint blockade. Cell Discov. 3, 17004.
Zhao, X., Rong, L., Zhao, X., Li, X., Liu, X., Deng, J., Wu, H., Xu, X., Erben, U., Wu, P., Syrbe, U., Sieper, J., Qin, Z., 2012. TNF signaling drives myeloid-derived suppressor cell accumulation. J. Clin. Invest. 122, 4094–4104.
Zhao, W.H., Liu, J., Wang, B.Y., Chen, Y.X., Cao, X.M., Yang, Y., Zhang, Y.L., Wang, F.X., Zhang, P.Y., Lei, B., Gu, L.F., Wang, J.L., Yang, N., Zhang, R., Zhang, H., Shen, Y., Bai, J., Xu, Y., Wang, X.G., Zhang, R.L., Wei, L.L., Li, Z.F., Li, Z.Z., Geng, Y., He, Q., Zhuang, Q.C., Fan, X.H., He, A.L., Zhang, W.G., 2018. A phase 1, open-label study of LCAR-B38M, a chimeric antigen receptor T cell therapy directed against B cell maturation antigen, in patients with relapsed or refractory multiple myeloma. J. Hematol. Oncol. 11, 141.
Zhao, J., Chen, X., Herjan, T., Li, X., 2020. The role of interleukin-17 in tumor development and progression. J. Exp. Med. 217.
Zhu, L., Yang, X., Zhong, D., Xie, S., Shi, W., Li, Y., Hou, X., HuaYao, Zhou, H., Zhao, M., Ding, Z., Zhao, X., Mo, F., Yin, S., Liu, A., Lu, X., 2020. Single-domain antibody-based TCR-like CAR-T: a potential cancer therapy. J. Immunol. Res. 2020, 2454907.

FSC
www.fsc.org
MIX
Papier | Fördert
gute Waldnutzung
FSC® C083411

Kolibri 360 GmbH
Ferdinand-Jühlke-Straße 7
99095 Erfurt, Deutschland
produktsicherheit@kolibri360.de